Student Solutions Manual

Introduction to General, Organic, and Biochemistry

TENTH EDITION

Frederick Bettelheim

William Brown
Beloit College

Mary Campbell
Mount Holyoke University

Shawn Farrell

Omar Torres
College of the Canyons

Prepared by

Mark Erickson
Hartwick College

Andrew Piefer
Hartwick College

BROOKS/COLE
CENGAGE Learning

Australia • Brazil • Japan • Korea • Mexico • Singapore • Spain • United Kingdom • United States

ISBN-13: 978-1-133-10910-5
ISBN-10: 1-133-10910-1

Brooks/Cole
20 Davis Drive
Belmont, CA 94002-3098
USA

Cengage Learning is a leading provider of customized learning solutions with office locations around the globe, including Singapore, the United Kingdom, Australia, Mexico, Brazil, and Japan. Locate your local office at: **www.cengage.com/global**

Cengage Learning products are represented in Canada by Nelson Education, Ltd.

To learn more about Brooks/Cole, visit **www.cengage.com/brookscole**

Purchase any of our products at your local college store or at our preferred online store **www.cengagebrain.com**

Printed in the United States of America
1 2 3 4 5 6 7 15 14 13 12 11

TABLE OF CONTENTS

Chapter 1: Matter, Energy, and Measurement

1.1 Multiplication: (a) 4.69×10^5 (b) 2.8×10^{-15}
  Division: (a) 1.94×10^{18} (b) 1.37×10^5

1.2 (a) $^\circ F = \dfrac{9}{5} \, ^\circ C + 32 = \dfrac{9}{5} \, 64.0^\circ C + 32 = 147^\circ F$

  (b) $^\circ C = \dfrac{5}{9} (^\circ F - 32) = \dfrac{5}{9} (47^\circ F - 32) = 8.3^\circ C$

1.3 $8.55 \text{ mi} \left(\dfrac{1.609 \text{ km}}{1 \text{ mi}} \right) = 13.8 \text{ km}$

1.4 $\dfrac{332 \text{ m}}{\text{s}} \left(\dfrac{1 \text{ km}}{1000 \text{ m}} \right) \left(\dfrac{1 \text{ mi}}{1.609 \text{ km}} \right) \left(\dfrac{60 \text{ s}}{1 \text{ min}} \right) \left(\dfrac{60 \text{ min}}{1 \text{ hr}} \right) = 743 \text{ mi/hr}$

1.5 $\dfrac{50. \text{ mL sol}}{\text{hr}} \left(\dfrac{1.5 \text{ g antibiotic}}{1000 \text{ mL sol}} \right) \left(\dfrac{1000 \text{ mg antibiotic}}{1 \text{ g antibiotic}} \right) \left(\dfrac{1 \text{ hr}}{60 \text{ min}} \right) = 1.3 \text{ mg antibiotic/min}$

1.6 Mass of $Ti = 17.3 \text{ mL} \left(\dfrac{4.54 \text{ g Ti}}{1 \text{ mL}} \right) = 78.5 \text{ g Ti}$

1.7 $d = m/V = \dfrac{56.8 \text{ g}}{23.4 \text{ mL}} = 2.43 \text{ g/mL}$

1.8 The specific gravity of a substance is the density of this substance divided by the density of water, which is 1.00 g/mL.

  $\text{Specific gravity} = 1.016 = \dfrac{d}{1.00 \text{ g/mL}}$

  $d = 1.016 \text{ g/mL}$

1.9 $\text{Amount of heat} = SH \times m \times (T_2 - T_1) = \dfrac{1.0 \text{ cal}}{\text{g} \cdot ^\circ C} (731 \text{ g})(74 - 8)\,^\circ C = 4.8 \times 10^4 \text{ cal}$

1.10 $(T_2 \text{-} T_1)°C = \dfrac{\text{heat}}{SH \times m} = \dfrac{230. \cancel{\text{cal}}}{(0.11\ \cancel{\text{cal}}/\cancel{g} \cdot °C)(100\ \cancel{g})} = 21°C$

100 g of iron may have been measured to one, two, or three significant figures (the zeros at the end of a number without a decimal point are called trailing zeros, are ambiguous and may or may not be significant). If 100 g implied two or three significant figures, T_2 will be 46° C (the number of significant figures is determined by the specific heat of iron, which is reported to two significant figures). One common way to show that trailing zeros are significant is to place a decimal point after the zeros, for example 100. g (three significant figures). Zeros between a decimal and a non-zero digit are significant figures. This method is used often in the textbook problems.

$(T_2 \text{-} T_1)°C = 21°C$

$T_2 = 21°C + 25°C = 46°C$

1.11 $SH = \dfrac{\text{Amount of heat}}{m \times (T_2 \text{-} T_1)} = \dfrac{88.2\ \text{cal}}{(13.4\ g)(176\text{-}23)°C} = 0.0430\ \text{cal}/g \cdot °C$

1.13 (a) Matter is anything that has mass and takes up space.
(b) Chemistry is the science that studies matter.

1.15 Dr. X's claim that the extract cured diabetes would be classified as a (c) hypothesis. No evidence had been provided to prove or disprove the claim.

1.17 (a) 3.51×10^{-1} (b) 6.021×10^2 (c) 1.28×10^{-4} (d) 6.28122×10^5

1.19 (a) 6.65×10^{17} (b) 1.2×10^1 (c) 3.9×10^{-16} (d) 3.5×10^{-23}

1.21 (a) 1.3×10^5 (b) 9.40×10^4 (c) 5.139×10^{-3}

1.25 (a) 2   (b) 5   (c) 5   (d) 5
(e) 3 or 5, depending on whether the zero's are significant
(f) 3   (g) 2

1.27 (a) 92   (b) 7.3   (c) 0.68   (d) 0.0032   (e) 5.9

1.29 (a) 1.53   (b) 2.2   (c) 4.8×10^{-4}

1.31 20 kiloseconds may have been measured to one or two significant figures (the zeros at the end of a number without a decimal point are ambiguous and may or may not be significant). If 20 kiloseconds implied two significant figures, the answers are rounded to two significant figures:

$$20 \text{ ks} \left(\frac{1000 \text{ s}}{1 \text{ ks}} \right) \left(\frac{1 \text{ min}}{60 \text{ s}} \right) = 3.3 \times 10^2 \text{ min}$$

$$20 \text{ ks} \left(\frac{1000 \text{ s}}{1 \text{ ks}} \right) \left(\frac{1 \text{ min}}{60 \text{ s}} \right) \left(\frac{1 \text{ hr}}{60 \text{ min}} \right) = 5.6 \text{ hr}$$

1.33 (a) 20 mm   (b) 1 inch   (c) 1 mile

1.35 (b) Weight would change slightly. Mass is independent of location, but weight is a force exerted by a body influenced by gravity. The influence of the Earth's gravity decreases with increasing distance from sea level.

1.37 Temperature conversions: $°F = \dfrac{9}{5} °C + 32$ and $K = 273 + °C$

(a) $°F = \dfrac{9}{5} 25°C + 32 = \underline{77°F}$ and $K = 273 + 25°C = \underline{298 \text{ K}}$

(b) $°F = \dfrac{9}{5} 40°C + 32 = \underline{104°F}$ and $K = 273 + 40°C = \underline{313 \text{ K}}$

(c) $°F = \dfrac{9}{5} 250°C + 32 = \underline{482°F}$ and $K = 273 + 250°C = \underline{523 \text{ K}}$

(d) $°F = \dfrac{9}{5} (-273)°C + 32 = \underline{-459°F}$ and $K = 273 + (-273)°C = \underline{0 \text{ K}}$

1.39 Metric unit conversions:

(a) $96.4 \; \cancel{mL} \left(\dfrac{1 \; L}{1000 \; \cancel{mL}} \right) = 0.0964 \; L$

(b) $275 \; \cancel{mm} \left(\dfrac{1 \; cm}{10 \; \cancel{mm}} \right) = 27.5 \; cm$

(c) $45.7 \; \cancel{kg} \left(\dfrac{1000 \; g}{1 \; \cancel{kg}} \right) = 4.57 \times 10^4 \; g$

(d) $475 \; \cancel{cm} \left(\dfrac{1 \; m}{100 \; \cancel{cm}} \right) = 4.75 \; m$

(e) $21.64 \; \cancel{cc} \left(\dfrac{1 \; mL}{1 \; \cancel{cc}} \right) = 21.64 \; mL$

(f) $3.29 \; \cancel{L} \left(\dfrac{1000 \; cc}{1 \; \cancel{L}} \right) = 3.29 \times 10^3 \; cc$

(g) $0.044 \; \cancel{L} \left(\dfrac{1000 \; mL}{1 \; \cancel{L}} \right) = 44 \; mL$

(h) $711 \; \cancel{g} \left(\dfrac{1 \; kg}{1000 \; \cancel{g}} \right) = 0.711 \; kg$

(i) $63.7 \; \cancel{mL} \left(\dfrac{1 \; cc}{1 \; \cancel{mL}} \right) = 63.7 \; cc$

(j) $0.073 \; \cancel{kg} \left(\dfrac{1000 \; g}{1 \; \cancel{kg}} \right) \left(\dfrac{1000 \; mg}{1 \; g} \right) = 7.3 \times 10^4 \; mg$

(k) $83.4 \; \cancel{m} \left(\dfrac{1000 \; mm}{1 \; \cancel{m}} \right) = 8.34 \times 10^4 \; mm$

(l) $361 \; \cancel{mg} \left(\dfrac{1 \; g}{1000 \; \cancel{mg}} \right) = 0.361 \; g$

1.41 $4.00 \; \cancel{gal} \left(\dfrac{4 \; \cancel{qt}}{1 \; \cancel{gal}} \right) \left(\dfrac{32 \; oz}{1 \; \cancel{qt}} \right) = 512 \; oz$

1.43 $\text{Speed limit} = \left(\dfrac{80 \; \cancel{km}}{hr} \right) \left(\dfrac{1 \; mi}{1.609 \; \cancel{km}} \right) = 50 \; mph$

1.45 $70. \; \cancel{lb} \; \text{child} \left(\dfrac{1 \; \cancel{kg}}{2.205 \; \cancel{lb}} \right) \left(\dfrac{10. \; \cancel{mg \; Ace}}{1 \; \cancel{kg}} \right) \left(\dfrac{1 \; \text{tablet Ace}}{80. \; \cancel{mg \; Ace}} \right) = 4.0 \; \text{tablets}$

1.47 $8.0 \; \cancel{hr} \left(\dfrac{400. \; \cancel{mL}}{24 \; \cancel{hr}} \right) \left(\dfrac{120. \; mg \; drug}{1000. \; \cancel{mL}} \right) = 16 \; mg \; drug$

1.49 $\left(\dfrac{1100 \; \cancel{units \; hep}}{hr} \right) \left(\dfrac{1 \; \cancel{L \; IV \; sol}}{26,000 \; \cancel{units \; hep}} \right) \left(\dfrac{1000 \; \cancel{mL \; IV \; sol}}{1 \; \cancel{L \; IV \; sol}} \right) \left(\dfrac{1 \; mL \; IV \; sol}{1 \; \cancel{cc \; IV \; sol}} \right) = \dfrac{42 \; cc \; IV \; sol}{hr}$

1.51 $139 \text{ lb} \left(\dfrac{1 \text{ kg}}{2.205 \text{ lb}} \right) \left(\dfrac{50. \text{ mL drug}}{1 \text{ kg}} \right) \left(\dfrac{20. \text{ gtts}}{1 \text{ mL drug}} \right) \left(\dfrac{1 \text{ min}}{150 \text{ gtts}} \right) = 4.2 \times 10^2 \text{ min}$

1.53 Liquids and solids have definite volumes.

1.55 Melting is a physical change, not a chemical change; therefore, when a substance melts from a solid to a liquid, its chemical nature does not change.

1.57 Manganese (d = 7.21 g/mL) is denser than the liquid (d = 2.15 g/mL); therefore it will sink. Sodium acetate (d = 1.528 g/mL) is less dense than the liquid; therefore it will float on the liquid. Calcium chloride (d = 2.15 g/mL) has a density equal to that of the liquid; therefore it will stay in the middle of the liquid.

1.59 $4 \text{ mg valium} \left(\dfrac{1 \text{ mL sol}}{5 \text{ mg valium}} \right) = 0.8 \text{ mL solution injected}$

1.61 Water (d = 1.0 g/cc) will be the top layer in the mixture because its density is lower than dichloromethane (d = 1.33 g/cc).

1.63 One should raise the temperature of water to 4°C. Water reaches its maximum density at 4°C. During this temperature change, the density of the crystals decreases, while the density of water increases. This brings the less dense crystals to the surface of the more dense water.

1.65 While driving your car, the car's <u>kinetic energy</u> (energy of motion) is converted by the alternator to electrical energy, which charges the battery, storing <u>potential energy</u>.

1.67 $SH_{unknown} = \dfrac{Heat}{m \times (T_2 - T_1)}$

$SH_{unknown} = \dfrac{2750 \text{ cal}}{168 \text{ g} (74° - 26°)°C} = 0.34 \text{ cal} / \text{g} \cdot °C$

1.69 Assuming that the 180 lb man is measured to three significant figures:

$\text{Drug dose}_{135lb} = 135 \text{ lb man} \left(\dfrac{445 \text{ mg drug}}{180 \text{ lb man}} \right) = 334 \text{ mg drug}$

1.71 The body first reacts to hypothermia by shivering. Further temperature lowering results in unconsciousness and later, followed by death.

1.72 Shivering generates heat for the body through muscular action.

1.73 Methanol would make a more effective cold compress because its higher specific heat allows it to retain the heat longer per mass unit.

1.75 $d_{brain} = \dfrac{1.0 \text{ lb}}{620 \text{ mL}}\left(\dfrac{453.6 \text{ g}}{1 \text{ lb}}\right) = 0.73 \text{ g/mL}$

$\text{Specific gravity}_{brain} = \dfrac{d_{brain}}{d_{H_2O}} = \dfrac{0.73 \text{ g/mL}}{1 \text{ g/mL}} = 0.73$

1.77 (a) Potential energy   (b) Kinetic energy   (c) Potential energy
(d) Kinetic energy   (e) Kinetic energy

1.79 Convert European car's fuel efficiency of 22 km/L into mi/gal, then compare:

$\text{Fuel efficiency}_{European} = \dfrac{22 \text{ km}}{L}\left(\dfrac{1 \text{ mi}}{1.609 \text{ km}}\right)\left(\dfrac{3.785 \text{ L}}{1 \text{ gal}}\right) = 52 \text{ mi/gal}$

The European car is more fuel efficient by 22 miles per gallon

1.81 Shivering is a form of kinetic energy that generates heat by way of frictional effects within body tissues.

1.83 Convert each quantity into a common unit (grams): (a) is the largest and (d) is the smallest.
(a) 41 g
(b) $3 \times 10^3 \text{ mg}\left(\dfrac{1 \text{ g}}{1000 \text{ mg}}\right) = 3 \text{ g}$
(c) $8.2 \times 10^6 \text{ µg}\left(\dfrac{1 \text{ g}}{10^6 \text{ µg}}\right) = 8.2 \text{ g}$
(d) $4.1310 \times 10^{-8} \text{ kg}\left(\dfrac{1000 \text{ g}}{1 \text{ kg}}\right) = 4.1310 \times 10^{-5} \text{ g}$

1.85 $\text{Travel time} = 1490. \text{ mi}\left(\dfrac{1.609 \text{ km}}{1 \text{ mi}}\right)\left(\dfrac{1 \text{ hr}}{220. \text{ km}}\right) = 10.9 \text{ hr}$

1.87 SH (water) = 1.000 cal/g · °C = 4.184 J/g · °C

SH (heavy water) = 4.217 J/g °C

Heat = SH x m x ΔT: According to the equation, the heat required to raise the temperature of a substance is directly proportional to the specific heat of that substance. Heavy water, having the higher specific heat, will require more energy to heat 10 g by 10°C.

1.89 1.00 mL of butter = 0.860 g and 1.00 mL of sand = 2.28 g

(a) $d_{mixture} = \dfrac{3.14 \text{ g mixture}}{2.00 \text{ mL}} = 1.57 \text{ g/mL}$

(b) First calculate the volumes of sand and butter, which totals 1.60 mL

$$V_{sand} = 1.00 \text{ g sand}\left(\frac{1.00 \text{ mL sand}}{2.28 \text{ g sand}}\right) = 0.439 \text{ mL}$$

$$V_{butter} = 1.00 \text{ g butter}\left(\frac{1.00 \text{ mL butter}}{0.860 \text{ g butter}}\right) = 1.16 \text{ mL}$$

$$d_{mixture} = \frac{2.00 \text{ g}}{1.60 \text{ mL}} = 1.25 \text{ g/mL}$$

1.91 The final answer will be reported to two significant digits because the temperature is reported in two significant figures (the least accurate of the quantities used in the calculation)

$$SH_{unk} = \frac{\text{Heat}}{m \times (T_2 - T_1)} = \frac{3.200 \text{ kcal}}{(92.15 \text{ g})(45°C)} = 7.7 \times 10^{-4} \text{ kcal}/\text{g} \cdot °C = 0.77 \text{ cal}/\text{g} \cdot °C$$

1.93 Assume that 20 °C represents two significant figures.

$$T_2 = \frac{\text{Heat}}{SH \times m} + T_1$$

$$T_2 = \frac{60.0 \text{ J}}{(10.0 \text{ g})(1.339 \text{ J/g} \cdot °C)} + 20 °C = 24 °C$$

1.95 Item (b) has three significant figures where item (a) has only one significant figure.

1.97 The estimate that 95 km is equivalent to 150 miles was incorrect. The correct conversion shows that 95 km is equivalent to 59 miles:

$$95 \text{ km} \left(\frac{1 \text{ mi}}{1.609 \text{ km}} \right) = 59 \text{ mi}$$

What most likely happened in the error was inverting the conversion factor:

$$95 \text{ km} \left(\frac{1609 \text{ km}}{1 \text{ mi}} \right) = 150 \frac{\text{km}^2}{\text{mi}}$$

Notice that in the error, the units do not cancel to give the value in miles, thus upon closer inspection, makes no sense. The best way to avoid this error is to always write down the units of each number used in the calculation and to treat them mathematically, canceling them when possible to give the answer in the simplest units.

1.99 Solar energy (light energy) is converted to chemical energy (potential energy) in the sugars through photosynthesis.

1.101 First, convert to degrees Fahrenheit:

$$°F = \frac{9}{5} °C + 32 = \frac{9}{5} 30° C + 32 = 86° F \quad (90° \text{ F to one significant figure})$$

A T-shirt will be more comfortable under these conditions.

1.103 As water inside a cell freezes, it expands causing the cell wall to burst when the cell's maximum volume is exceeded.

1.105 Heat = SH × m × ΔT = $(1.00 \text{ cal g}^{-1}°C^{-1})(2000. \text{ g})(4.85°C) = 9.70 \times 10^3$ cal

1.107 The folk medicine will be extracted and separated into pure materials. Each pure material will be tested for biological activity. The pure materials that are identified as biologically active will have their chemical structures determined, and often, a chemical synthesis developed for mass production.

1.109 Heat = $\left(\frac{9.70 \times 10^3 \text{ cal}}{2 \text{ kg}} \right) = 5 \times 10^3$ cal/kg (rounded to one significant figure)

<u>1.111</u> (a) $\left(\dfrac{100.\ \text{mg drug}}{\text{hr}}\right)\left(\dfrac{1\ \text{g drug}}{1000\ \text{mg drug}}\right)\left(\dfrac{1000\ \text{mL sol}}{5.0\ \text{g drug}}\right) = 20.\ \text{mL/hr}$

(b) $\left(\dfrac{20.\ \text{mL sol}}{\text{hr}}\right)\left(\dfrac{1\ \text{hr}}{60\ \text{min}}\right)\left(\dfrac{15\ \text{gtts}}{1\ \text{mL}}\right) = 5.0\ \text{gtts/min}$

The set drip rate of 10. gtts/min is too fast. The drip rate should be reset to 5.0 gtts.

Chapter 2: Atoms

2.1 (a) $NaClO_3$ (b) AlF_3

2.2 (a) The mass number is 15 + 16 = 31.
    (b) The mass number is 86 + 136 = 222.

2.3 (a) The element has 15 protons, making it phosphorus (P); its symbol is $^{31}_{15}P$.

    (b) The element has 86 protons, making it radon (Rn); its symbol is $^{222}_{86}Rn$.

2.4 (a) The atomic number of mercury (Hg) is 80; that of lead (Pb) is 82.
    (b) An atom of Hg has 80 protons; an atom of Pb has 82 protons.
    (c) The mass number of this isotope of Hg is 80 + 120 = 200; the mass number for this
        isotope of Pb is 82 + 120 = 202.
    (d) The symbols of these isotopes are $^{200}_{80}Hg$ and $^{202}_{82}Pb$.

2.5 The atomic number of iodine (I) is 53. The number of neutrons in each isotope is 125 - 53
    = 72 for iodine-125 and 131 - 53 = 78 for iodine-131. The symbols for these two isotopes
    are $^{125}_{53}I$ and $^{131}_{53}I$.

2.6 The atomic weight is 6.941 amu, which is nearer to 7 amu than 6 amu. Therefore, lithium-
    7 is the more abundant isotope. The relative abundances for these two isotopes are 92.50
    percent for lithium-7 and 7.50 percent for lithium-6.

2.7 This element has 13 electrons and, therefore, 13 protons. The element with atomic number
    13 is Aluminum (Al).

$$\dot{Al}:$$

2.8 Both Democritus and Dalton believed that matter was composed of tiny indivisible
    particles referred to as atoms. The major difference between Democritus and Dalton is that
    Dalton based his theory on evidence rather than belief.

2.9 (b), (c), (d), (f), (g), (h), and (k): True
    (a) False: matter is divided into pure substances and mixtures.
    (e) False: mixtures can be separated into their component pure substances.
    (i) False: technicium, promethium, and all of the elements beyond uranium are man made.
    (j) False: H, O, C, N, Ca, and P are the six most important elements in the human body.
    (l) False: The combining ratio is based on a ratio of atoms, not a ratio of masses.

2.10 (a) Oxygen - an element   (b) Table salt - a compound
     (c) Sea water - a mixture   (d) Wine - a mixture
     (e) Air - a mixture   (f) Silver - an element
     (g) Diamond - an element   (h) A pebble - a mixture
     (i) Gasoline - a mixture   (j) Milk - a mixture
     (k) Carbon dioxide - a compound   (l) Bronze - a mixture

2.11 (a) Oxygen   (b) Lead   (c) Calcium   (d) Sodium   (e) Carbon   (f) Titanium
     (g) Sulfur   (h) Iron   (i) Hydrogen   (j) Potassium   (k) Silver   (l) Gold

2.13 Given here is the element, its symbol, and its atomic number:
     (a) Americium (Am, 95)   (b) Berkelium (Bk, 97)   (c) Californium (Cf, 98)
     (d) Dubnium (Db, 105)   (e) Europium (Eu, 63)   (f) Francium (Fr, 87)
     (g) Gallium (Ga, 31)   (h) Germanium (Ge, 32)   (i) Hafnium (Hf, 72)
     (j) Hassium (Hs, 108)   (k) Holmium (Ho, 67)   (l) Lutetium (Lu, 71)
     (m) Magnesium (Mg, 12)   (n) Polonium (Po, 84)   (o) Rhenium Re, 75)
     (p) Ruthenium (Ru, 44)   (q) Scandium (Sc, 21)   (r) Strontium (Sr, 38)
     (s) Ytterbium (Yb, 70), Terbium (Tb, 65), and Yttrium (Y, 39)
     (t) Thulium (Tm, 69)

2.15 (a) K_2O (b) Na_3PO_4 (c) $LiNO_3$

2.17 (a) The law of conservation of mass states that matter can be neither created nor destroyed. Dalton's theory explains this because if all matter is made up of indestructible atoms, then any chemical reaction just changes the attachments between atoms and does not destroy the atoms themselves.
     (b) The law of constant composition states that any compound is always made up of elements in the same proportion by mass. Dalton's theory explains this because molecules consist of tightly bound groups of atoms, each of which has a particular mass. Therefore, each element in a compound always constitutes a fixed proportion of the total mass.

2.19 No. CO and CO_2 are different compounds, and each obeys the law of constant composition for that particular compound.

© 2013 Cengage Learning. All Rights Reserved. May not be scanned, copied or duplicated, or posted to a publicly accessible website, in whole or in part.

2.21 (b), (c), (e), (f), (g), (h), (k), (m), (o), (p), (q), and (s): True
    (a) False: electrons and protons have equal, but opposite charges. Electrons are much lighter in mass than protons.
    (d) False: 1 amu = 1.6605 x 10^{-24} grams.
    (i) False: opposite charges attract each other.
    (j) False: the size of the atom includes the space occupied by the electrons. The nucleus is a small fraction of the size of the atom.
    (l) False: The mass number is the number of protons and neutrons.
    (n) False: ^{1}H has no neutrons, ^{2}H has one neutron, and ^{3}H has two neutrons.
    (r) False: atomic weights are averages of the known isotopes.

    (t) False: Density is mass/volume.

2.23 The statement is true in the sense that the number of protons (the atomic number) determines the identity of the element.

2.25 (a) The element with 22 protons is titanium (Ti).
    (b) The element with 76 protons is osmium (Os).
    (c) The element with 34 protons is selenium (Se).
    (d) The element with 94 protons is plutonium (Pu).

2.27 Each would still be the same element because the number of protons has not changed.
    (a) The element is scandium (Sc), and its symbol is $^{47}_{21}$Sc .

    (b) The element is titanium (Ti), and its symbol is $^{50}_{22}$Ti .

    (c) The element is silver (Ag), and its symbol is $^{109}_{47}$Ag .

    (d) The element is thorium (Th) and its symbol is $^{248}_{90}$Th .

    (e) The element is argon (Ar) and its symbol is $^{38}_{18}$Ar .

2.29 Radon (Rn) has an atomic number of 86, so each isotope has 86 protons. The number of neutrons is mass number - atomic number.
    (a) Radon-210 has 210 - 86 = 124 neutrons
    (b) Radon-218 has 218 - 86 = 132 neutrons
    (c) Radon-222 has 222 - 86 = 136 neutrons

2.31 Two more neutrons: tin-120
    Three more neutrons: tin-121
    Four more neutrons: tin 124

2.33 (a) An ion is an atom or group of bonded atoms with an unequal number of protons and electrons.

(b) Isotopes are atoms with the same number of protons in their nuclei but a different number of neutrons.

2.35 Rounded to four significant figures, the calculated value is 12.01 amu. The value given in the Periodic Table is 12.011 amu.

$$\left(\frac{98.90}{100} \times 12.000 \text{ amu}\right) + \left(\frac{1.10}{100} \times 13.003 \text{ amu}\right) = 12.01 \text{ amu}$$

2.37 Carbon-11 has 6 protons, 6 electrons, and 5 neutrons.

2.39 Americium-241 (Am) has atomic number 95. This isotope has 95 protons, 95 electrons and 241 - 95 = 146 neutrons.

2.41 (a), (f), (g), (h), and (i): True

(b) False: main group elements go from groups 1A through 8A

(c) False: very roughly, the metalloids exist in a diagonal, starting in the lower right corner, moving up to upper middle of the Periodic Table. Nonmetals exist above the diagonal, metals below it.

(d) False: there are more metals than nonmetals

(e) False: horizonal rows are called periods.

2.43 (a) Groups 2A, 3B, 4B, 5B, 6B, 7B, 8B, 1B, and 2B contain only metals. Note that Group 1A contains one nonmetal, hydrogen.

(b) No group contains only metalloids.

(c) Only Groups 7A and 8A contain only nonmetals.

2.45 In the Periodic Table, elements of the same group should have similar properties: As, N, and P; I, and F; Ne and He; Mg, Ca, and Ba; K and Li.

2.47 (a) Aluminum > silicon   (b) Arsenic > phosphorus

(c) Gallium > germanium   (d) Gallium > aluminum

2.49 (a), (b), (c), (e), (g), (h), (j), (k), (l), (m), (n), (q), (r), and (s): True
  (d) False: principal energy level 1 contains a maximum of two electrons, the principal
    energy level 2 contains a maximum of eight electrons, the principal energy
    level 3 contains a maximum of 18 electrons, and the principal energy level 4
    contains a maximum of 32 electrons.
  (f) False: a $2s$ electron is easier to remove than a $1s$ electron because it is further away
    from the influence of the positively charged nucleus.
  (i) False: The three $2p$ orbitals are arranged perpendicular to each other.
  (o) False: paired electrons have spins in the opposite direction.
  (p) False: each box represents an orbital and each orbital can accommodate two electrons.
    When an orbital is completely filled, the electrons must be paired with spins in
    the opposite direction.
  (t) False: group 6A elements have two unpaired electrons.

2.51 The group number indicates the number of electrons in the valence shell of the element.

2.53 (a) Li(3): $1s^2 2s^1$ (b) Ne(10): $1s^2 2s^2 2p^6$ (c) Be(4): $1s^2 2s^2$
  (d) C(6): $1s^2 2s^2 2p^2$ (e) Mg(12): $1s^2 2s^2 2p^6 3s^2$

2.55 (a) He(2): $1s^2$ (b) Na(11): $1s^2 2s^2 2p^6 3s^1$ (c) Cl(17): $1s^2 2s^2 2p^6 3s^2 3p^5$
  (d) P(15): $1s^2 2s^2 2p^6 3s^2 3p^3$ (e) H(1): $1s^1$

2.57 In (a), (b), and (c); the outer-shell electron configurations are the same. The only
  difference is the number of the valence shell being filled.

2.59 The element might be in Group 2A, all of which have two valence electrons. It might also
  be helium.

2.61 (a), (b), (c), (f), and (h): True
  (d) False: helium is a group 8A element and has two valence electrons.
  (e) False: the period number has nothing to do with the number of valence electrons.
  (g) False: period 3 has eight elements.

2.63 (a), (b), (c), (d), and (f): True
  (e) False: ionization energy decreases going from top to bottom within a column of the
    Periodic Table.

2.65 (a) The additional electron in the valence shell exceeds the positive charge in the nucleus causing the electrons in the valence shell to be held less tightly, thus increasing the atomic radius. Also, the additional electron in the valence shell introduces new repulsions causing the electron cloud to expand.

(b) With the loss of a valence electron, the positive charge in the nucleus exceeds the collective negative charge of the electrons, causing the valence electrons to held more tightly, thus decreasing the atomic radius. Also, one less electron in the valence shell reduces the electron-electron repulsions causing the electron cloud to contract.

2.67 Electron configurations of atoms with either half-filled or filled shells are especially stable. The favorable first ionization of oxygen leads to an electron configuration with a stable half-filled $2p$ subshell. The first ionization of nitrogen is more difficult because it loses an electron from an already stable half-filled $2p$ subshell to a less stable electron configuration of a partially filled subshell.

$$\text{O } 1s^2 2s^2 2p_x^2 2p_y^1 2p_z^1 \rightarrow \text{O}^+ \ 1s^2 2s^2 2p_x^1 2p_y^1 2p_z^1 + 1 \ e^- \ (\text{half filled } 2p \text{ subshell})$$

$$\text{N } 1s^2 2s^2 2p_x^1 2p_y^1 2p_z^1 \rightarrow \text{N}^+ \ 1s^2 2s^2 2p_x^1 2p_y^1 2p_z^0 + 1 \ e^-$$

2.69 Sulfur and iron are essential components of proteins, and calcium is a major component of bones and teeth.

2.71 Calcium is an essential element in human bones and teeth. Because strontium behaves chemically much like calcium, strontium-90 gets into our bones and teeth and gives off radioactivity for many years directly into our bodies.

2.73 Copper can be made harder by hammering it.

2.75 (a) Metals   (b) Nonmetals   (c) Metals
    (d) Nonmetals   (e) Metals   (f) Metals

2.77 The atomic radius decreases going from left to right across a period in the Period Table. Atomic radius increases going down a group (column) in the Periodic Table.
(a) Radium (Ra)   (d) Lithium (Li)
(b) Beryllium (Be)   (e) Fluorene (F)
(c) Neon (Ne)   (f) Astatine (At)

2.79 (a) Phosporous-32 has 15 protons, 15 electrons, and 32 - 15 = 17 neutrons.
(b) Molybdenum-98 has 42 protons, 42 electrons, and 98 - 42 = 56 neutrons.
(c) Calcium-44 has 20 protons, 20 electrons, and 44 - 20 = 24 neutrons.
(d) Hydrogen-3 has 1 proton, 1 electron, and 3 - 1 = 2 neutrons.
(e) Gadolinium-158 has 64 protons, 64 electrons, and 158 - 64 = 94 neutrons.
(f) Bismuth-212 has 83 protons, 83 electrons, and 212 - 83 = 129 neutrons.

2.81 For elements with atomic numbers less than that of iron (Fe), the number of neutrons is close to the number of protons. Elements with atomic numbers greater than that of iron have more neutrons than protons. Therefore, heavy elements have more neutrons than protons.

2.83 Rounded to four significant figures, the atomic weight of naturally occurring boron is 10.81. The value given in the Periodic Table is 10.811.

$$\left(\frac{19.9}{100} \times 10.013 \text{ amu}\right) + \left(\frac{80.1}{100} \times 11.009 \text{ amu}\right) = 10.8 \text{ amu}$$

2.85 It would take 6.0×10^{21} protons to equal the mass of a grain of salt.

$$\frac{1.0 \times 10^{-2} \text{ g NaCl}}{1.67 \times 10^{-24} \text{ g/proton}} = 6.0 \times 10^{21} \text{ protons}$$

2.87 The atomic number of this element is 54, which means that it is xenon (Xe). This isotope of xenon has 54 protons, 54 electrons, and $131 - 54 = 77$ neutrons.

2.89 (a) Ionization energy generally decreases down a column in the Periodic Table, so the ionization energy of element 117 should be less than that of At(85).
(b) Ionization energy generally increases from left to right across a row in the Periodic Table, so ionization energy of element 117 should be greater than Ra(88).

2.91 (a) The first electron is removed from the 2s orbital. The removal of each subsequent electron requires more energy because, after the first electron is removed, each subsequent electron is removed from a positive ion, which strongly attracts the remaining electrons. The second ionization energy is especially large because the electron is removed from the filled first principal energy level, meaning that it is removed from an ion that has the same electron configuration as helium.
(b) Li^+ ion has a smaller radius than Li because with the loss of a valence electron from Li, the positive charge in the nucleus exceeds the collective negative charge of the electrons, causing the valence electrons to be held more tightly, thus decreasing the atomic radius. Also, one less electron in the valence shell of Li^+ reduces the electron-electron repulsions causing the electron cloud to contract.

2.93 Increasing size: C < B < Al < Na
There are relatively small decreases in atomic radii going from left to right in the Periodic Table due to the increasingly stronger pull on the electrons because of the additional protons, therefore, the size of the atomic radii can be compared as C < B and Al < Na. Going to a higher numbered period in the Period Table results in much larger increase in atomic radii, therefore, both Na and Al are larger than C and B.

2.95 Going down Group 3A, we have boron, aluminum, gallium, indium, and thallium. The electron configuration of each 3A element tells us that the valence shell s orbitals are completely filled and that the valence shell p orbital has one electron.

2.97 The Ca^{3+} ion is not found in chemical compounds because calcium only needs to lose two electrons to achieve a noble gas electronic configuration.

2.99 Element 118 will be in Group 8A. Expect it to be a gas that forms either no compounds or form very few compounds.

Chapter 3: Chemical Bonds

3.1 (a) A magnesium (Mg) atom with two valence electrons, loses both electrons to form a
      Mg^{2+} ion with a Neon (Ne) electron configuration.
      (b) A sulfur (S) atom with six valence electrons gains two electrons to give a sulfide ion
      (S^{2-}) with an eight valence electron octet of an argon electron configuration.

3.2 Electronegativity increases from left to right in a period and from bottom to top in a
      column.
      (a) Li > K   (b) N > P   (c) C > Si

3.3 (a) KCl   (b) CaF_2 (c) Fe_2O_3

3.4 (a) Magnesium oxide   (b) Barium iodide   (c) Potassium chloride

3.5 (a) $MgCl_2$ (b) Al_2O_3 (c) LiI

3.6 (a) Systematic Name: iron (II) oxide; common name: ferrous oxide
      (b) Systematic Name: iron (III) oxide; common name: ferric oxide

3.7 (a) Potassium hydrogen phosphate   (b) Aluminum sulfate
      (c) Iron (II) carbonate or Ferrous carbonate

3.8

Bond	Difference in Electronegativity	Type of Bond
(a) S-H	$2.5 - 2.1 = 0.4$	Nonpolar covalent
(b) P-H	$2.1 - 2.1 = 0.0$	Nonpolar covalent
(c) C-F	$4.0 - 2.5 = 1.5$	Polar covalent
(d) C-Cl	$3.0 - 2.5 = 0.5$	Polar covalent

3.9 (a) C–N   (b) N–O   (c) C–Cl

3.10 Lewis structures:

(a) H-C-C-H   (b) H-C-Cl   (c) H-C≡N

3.11 Lewis structures:

(a) H-C-H   (b) C=C   (c) O=C=O   (d) H-C≡C-H

methane   ethene   carbon dioxide   acetylene

3.12 (a) Nitrogen dioxide   (b) Phosphorus tribromide
     (c) Sulfur dichloride   (d) Boron trifluoride

3.13 Contributing structures:

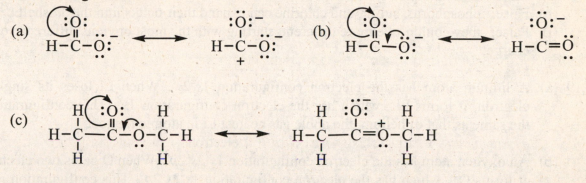

3.14 The structures in (a) represent contributing structures. They differ only in the distribution of valence electrons:

(a)

The structures in (b) do not represent contributing structures because carbon in the second structure exceeds its octet (10 electrons in five bonds to carbon).

(b)   five bonds to carbon

3.15 Given are three-dimensional structures showing all unshared electron pairs.

3.16

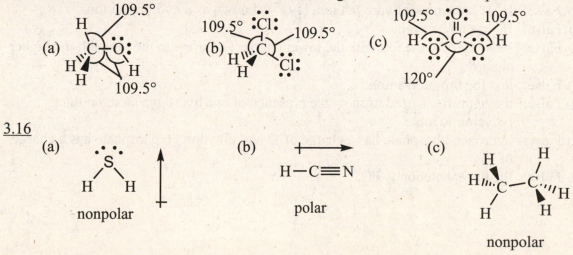

(a) nonpolar   (b) polar   (c) nonpolar

3.17 (b), (d), (e), and (g): True
   (a) False: it helps us understand the bonding patterns of the Group 1A-7A elements.
   (c) False: atoms that gain electrons become anions.
   (f) False: sodium typically forms positive ions by losing its $3s$ electron.
   (h) False: phosphorus, sulfur, and chlorine can expand their octets into the $3d$ shell.
   (i) False: they couldn't be more different, starting with the most obvious difference of charge.

3.19 (a) A lithium atom has the electron configuration $1s^2 2s^1$. When Li loses its single $2s$ electron, it forms Li^+, which has the electron configuration $1s^2$. This configuration is the same as that of helium, the noble gas nearest Li in atomic number.
$$Li: 1s^2 2s^1 \rightarrow Li^+: 1s^2 + 1 \text{ electron}$$
   (b) An oxygen atom has the electron configuration $1s^2 2s^2 2p^4$. When O gains two electrons, it forms O^{2-}, which has the electron configuration $1s^2 2s^2 2p^6$. This configuration is the same as that of neon, the noble gas nearest oxygen in atomic number.
$$O: 1s^2 2s^2 2p^4 + 2 \text{ electrons} \rightarrow O^{2-}: 1s^2 2s^2 2p^6 \text{ (complete octet)}$$

3.21 (a) Mg^{2+} (b) F^- (c) Al^{3+} (d) S^{2-} (e) K^+ (f) Br^-

3.23 The stable ions: (a) I^-, (c) Na^+, and (d) S^{2-} posses a filled octet. The other ions are unstable because they have unfilled shells or a charge that is too large.

3.25 Being of intermediate electronegativity, carbon and silicon are reluctant to accept electrons from a metal or lose electrons to a halogen to form ionic bonds. Instead, carbon and silicon share electrons in nonpolar covalent or polar covalent bonds.

3.27 (a), (d), (f), (h), (i), (m), and (n): True
   (b) False: H^+ is known as a hydrogen ion, H_3O^+ is known as a hydronium ion.
   (c) False: H^+ has no neutrons.
   (e) False: -ous refers to the ion with the lower charge, -ic refers to the ion with the higher charge.
   (g) False: it is the bromide anion.
   (j) False: the bi prefix is used to show the presence of one hydrogen atom on the polyatomic ion.
   (k) False: hydrogen phosphate has a charge of -2 and dihydrogen phosphate has a charge of -1.
   (l) False: the phosphate ion is PO_4^{3-}.

3.29 (a), (b), (c), (d), (f), (g), (h), (k), (m), and (n): True
    (e) False: ionic bonds usually form between elements on the far right and far left of the periodic table.
    (i) False: electronegativity is a periodic property.
    (j) False: electronegativity increases going from left to right in a period and decreases going down a group in the Periodic Table.
    (l) False: fluorine is the most electronegative atom and francium is the least electronegative.
    (o) False: the opposite is true.

3.31 Electronegativity generally increases in going from left to right across a row in the Periodic Table because the number of positive charges in the nucleus of each successive element in a row increases going from left to right. The increasing nuclear charge exerts a stronger and stronger pull on the valence electrons.

3.33 Electrons are shifted toward the more electronegative atom.
    (a) Cl   (b) O   (c) O   (d) Cl   (e) negligible   (f) negligible   (g) O

3.35 Use the difference in electronegativity to determine the character of the bond.
    (a) C-Cl (3.0 - 2.5 = 0.5); polar covalent   (b) C-Li (2.5 - 1.0 = 1.5); polar covalent
    (c) C-N (3.0 - 2.5 = 0.5); polar covalent

3.37 (a), (c), (d), and (e): True
    (b) False: ionic bonds form by the transfer of one or more valence electrons from the atom of lower electronegativity to the atom of higher electronegativity.
    (f) False: the correct formula is $Ca(OH)_2$.
    (g) False: the correct formula is Al_2S_3.
    (h) False: the correct formula is Fe_2O_3.
    (i) False: the correct formula is BaO.

3.39 (a) NaBr   (b) Na_2O (c) $AlCl_3$ (d) $BaCl_2$ (e) MgO

3.41 Sodium chloride in the solid state has a Na^+ cation surrounded by six Cl^- anions and each Cl^- anion surrounded by six Na^+ cations.

3.43 (a) $Fe(OH)_3$ (b) $BaCl_2$ (c) $Ca_3(PO_4)_2$ (d) $NaMnO_4$

3.45 (a) The formula, $(NH_4)_2PO_4$, is incorrect. The correct formula is: $(NH_4)_3PO_4$.
    (b) The formula, Ba_2CO_3, is incorrect. The correct formula is $BaCO_3$.
    (c) The formula for aluminum sulfide, Al_2S_3 is correct.
    (d) The formula for magnesium sulfide, MgS is correct.

3.47 (a), (c), (d), (g), and (i): True
     (b) False: the name includes no indication of the number of ions present.
     (e) False: the correct name is iron(III) oxide.
     (f) False: the correct name is iron(II) carbonate.
     (h) False: the correct name is potassium hydrogen phosphate.
     (j) False: the correct name is phosphorus trichloride.
     (k) False: the correct formula is $(NH_4)_2CO_3$.

3.49 The formula for potassium nitrite is KNO_2

3.51 (a) Na^+, Br^- (b) Fe^{2+}, SO_3^{2-} (c) Mg^{2+}, PO_4^{3-}
     (d) K^+, $H_2PO_4^-$ (e) Na^+, HCO_3^- (f) Ba^{2+}, NO_3^-

3.53 (a) KBr (b) CaO (c) HgO
     (d) $Cu_3(PO_4)_2$ (e) Li_2SO_4 (e) Fe_2S_3

3.55 (a), (d), (e), (f), (g), (i), (j), (l), and (n): True
     (b) False: they will form a nonpolar covalent bond.
     (c) False: a bond sharing two electrons is a single bond. Double bonds share four
            electrons.
     (h) False: the order should be reversed.
     (k) False: there are 14 electrons to the structure.
     (m) False: the structure should show eight electrons.

3.57 (a) A single bond results when one electron pair is shared between two atoms.
     (b) A double bond results when two electron pairs are shared between two atoms.
     (c) A triple bond results when three electron pairs are shared between two atoms.

3.59 Lewis structures for the following compounds:

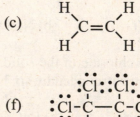

3.61 Total number of valence electrons for each compound:
     (a) NH_3 has 8   (b) C_3H_6 has 18   (c) $C_2H_4O_2$ has 24
     (d) C_2H_6O has 20   (e) CCl_4 has 32   (f) HNO_2 has 18
     (g) CCl_2F_2 has 32   (h) O_2 has 12

3.63 A bromine atom contains seven electrons in its valence shell. A bromine molecule contains two bromine atoms bonded by a single bond. A bromide ion is a bromine atom that has gained one electron in its valence shell; it has a complete octet and a charge of -1.

   (a) :Br·   (b) :Br—Br: (c) :Br:⁻

3.65 Hydrogen has the electron configuration $1s^1$. Hydrogen's valence shell has only a $1s$ orbital, which can hold only two electrons.

3.67 Nitrogen has five valence electrons. By sharing three more electrons with another atom(s), nitrogen can achieve the outer-shell electron configuration of neon, the noble gas nearest to it in atomic number. The three shared pairs of electrons may be in the form of three single bonds, one double bond and one single bond, or one triple bond. With any of these combinations, there is one unshared pair of electrons on nitrogen.

3.69 Oxygen has six valence electrons. By sharing two electrons with another atom(s), oxygen can achieve the outer-shell electron configuration of neon, the noble gas nearest to it in atomic number. The two shared pairs of electrons may be in the form of a double bond or two single bonds. With either of these configurations, there are two unshared pairs of electrons on the oxygen.

3.71 The O^{6+} cation has a large positive charge that is too concentrated for a small ion.

3.73 (a) BF_3 has six valence electrons around the boron, thus does not obey octet rule.
   (b) CF_2: does not obey the octet rule because carbon has 4 electrons around it.
   (c) BeF_2: does not obey the octet rule because Be has 4 electrons around it.
   (d) $H_2C=CH_2$ obeys the octet rule
   (e) CH_3 does not obey the octet rule because carbon has 6 electrons around it.
   (f) N_2 obeys the octet rule.
   (g) NO does not obey the octet rule. Lewis structures drawn for the compound show either a nitrogen atom or an oxygen atom with seven electrons.

3.75 (a) Sulfur dioxide   (b) Sulfur trioxide
   (c) Phosphorus trichloride   (d) Carbon disulfide

3.77 (a) A correct Lewis structure for ozone must contain 18 valence electrons.
(b) The two equivalent contributing structures for ozone:

(c) The curved arrows in (b) show two electron transfers between contributing structures.
(d) The structure of ozone will be determined by a combination of contributing structures. The contributing structures for ozone are equivalent, therefore, are equal contributors to the overall structure of ozone. Both structures have three areas of electron density; therefore VSEPR theory would predict an O-O-O bond angle of 120°.
(e) This Lewis structure involves a central oxygen that violates the octet rule (10 valence electrons).

3.79 (a), (c), (e), (f), (h), (i), (j), (l), and (m): True
(b) False: predicting structure must also consider non-bonding pairs of electrons.
(d) False: in the VSEPR theory, each double bond is considered one region of electron density.
(g) False: four regions of electron density result in bond angles of approximately 109.5°.
(k) False: H_2O has four regions of electron density, therefore bond angles of 109.5° are the result.

3.81 (a) H_2O has 8 valence electrons, and H_2O_2 has 14 valence electrons.

$$H-\overset{\cdot\cdot}{\underset{\cdot\cdot}{O}}-H \qquad H-\overset{\cdot\cdot}{\underset{\cdot\cdot}{O}}-\overset{\cdot\cdot}{\underset{\cdot\cdot}{O}}-H$$

(c) Predicted bond angles of 109.5° about each oxygen atom.

3.83 Shape of the each molecule and approximate bond angles about its central atom:
(a) Tetrahedral, 109.5°   (b) Pyramidal, 109.5°   (c) Tetrahedral, 109.5°
(d) Bent, 120°   (e) Trigonal planar, 120° (f) Tetrahedral, 109.5°
(g) Pyramidal, 109.5°   (h) Pyramidal, 109.5°

3.85 (a), (b), (d), (e), (f), (g), and (h): True
(c) False: if the dipole moments of polar bonds cancel each other by acting in equal, but opposite directions, the molecule will be nonpolar.

3.87 (a) Lewis structure for BF_3:

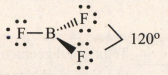

Trigonal planar

(b) According to the VSEPR theory, the bond angles around B are predicted to be 120°.
(c) Although the B-F bonds are polar, the sum of the individual B-F bonds in a trigonal planar geometry result in a nonpolar molecule.

3.89 No, it is not possible to have a polar molecule with nonpolar bonds. Molecular dipoles are the result of the sum of the direction and magnitude of each individual polar bond.

3.91 The individual C-Cl bond dipoles in CCl_4 act in equal, but opposite directions, canceling each other's effect on the molecular dipole.

3.93 Sodium iodide, NaI, and potassium iodide, KI, are used as iodide sources in table salt. Iodide is necessary for proper thyroid and thyroid hormone function.

3.95 Potassium permanganate is used as an external antiseptic.

3.97 Nitric oxide, NO, quickly oxidizes to nitrogen dioxide, NO_2, which then dissolves in rain to form nitric acid, HNO_3.

3.99 (a) $SiCl_4$ (b) PH_3 (c) H_2S

3.101 Putting together the bases of two square-based pyramids forms an octahedral geometry because it has eight faces.

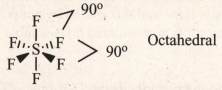

Octahedral

The VSEPR theory predicts an octahedral structure, which places the bonding pairs of electrons as far apart as possible, thereby minimizing electron pair repulsions.

3.103 Using the atomic radii for H (37 pm), O (66 pm), and S (104 pm), the O-H bond in H_2O can be estimated to be 103 pm and the S-H bond in H_2S can be estimated to be 141 pm.

3.105 (a) and (b): Bond angle predictions are approximate based on VSEPR theory.

Bent geometry with
H-O-C bond angles
of 109°.

Tetrahedral geometry with bond C-C-C, C-C-H,
and H-C-H angles of approximately 109°.

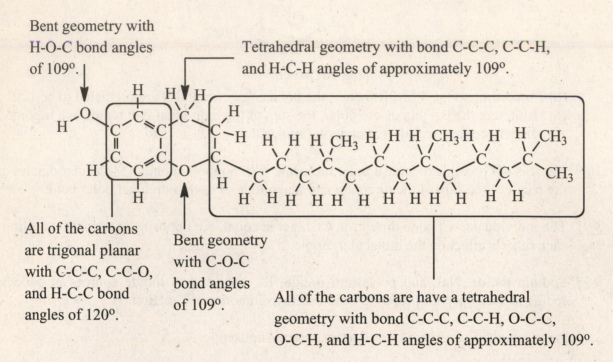

All of the carbons
are trigonal planar
with C-C-C, C-C-O,
and H-C-C bond
angles of 120°.

Bent geometry
with C-O-C
bond angles
of 109°.

All of the carbons are have a tetrahedral
geometry with bond C-C-C, C-C-H, O-C-C,
O-C-H, and H-C-H angles of approximately 109°.

(c) The O-H bond is the most polar bond in the vitamin E molecule. This is determined quantitatively by calculating the differences between the Pauling electronegativity values (Table 3.5) for the atoms in each bond, giving the O-H bond the largest electronegativity difference of 1.4.

(d) According to the definition outlined in the text, a molecule is polar if it has polar bonds and its centers of partial positive and negative charges are in different places within the molecule. Although vitamin E has a polar O-H bond and three polar C-O bonds, the molecule is large and dominated by nonpolar C-H and C-C bonds. Because of this, vitamin E is considered a fat-soluble vitamin, a characteristic of other non-polar molecules.

3.107 (a) The most polar bond in ephedrine is the O-H bond. This is determined quantitatively by calculating the differences between the Pauling electronegativity values (Table 3.5) for the atoms in each bond, giving the O-H bond the largest electronegativity difference of 1.4.

(b) According to the definition outlined in the text, a molecule is polar if it has polar bonds and its centers of partial positive and negative charges are in different places within the molecule. Ephedrine has two very polar O-H and N-H bonds, two polar C-N bonds, and a polar C-O bond in the right geometries. The presence of numerous polar bonds in a small molecule makes ephedrine a polar molecule.

3.109 The following structures illustrate are planar, the arrows showing the direction of the molecular dipole moment.

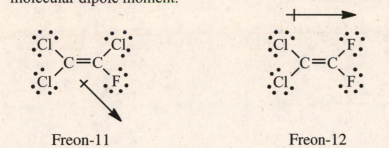

Freon-11   Freon-12

3.111 Zinc oxide, ZnO, is present in sun blocking agents and helps reflect sunlight away from the skin.

3.113 Lead(IV) oxide, PbO_2, and lead(IV) carbonate, $Pb(CO_3)_2$, were used as pigments in paint.

3.115 Fe(II) is utilized in over-the-counter iron supplements.

3.117 (a) $CaSO_3$ (b) $Ca(HSO_3)_2$ (c) $Ca(OH)_2$ (d) $CaHPO_4$

3.119 Perchloroethylene does possess four polar covalent C-Cl bonds, but is not a polar compound. The molecule lacks a dipole because the polar covalent C-Cl bond dipoles act in equal, but opposite directions.

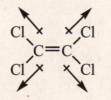

3.121 (a) Following is a Lewis structure for tetrafluoroethylene.
       (b) All bond angles are predicted to be 120°.
       (c) Tetrafluoroethylene has four polar covalent C-F bonds, but is a nonpolar molecule, because like in problem 3.119, the polar covalent C-F bond dipoles act in equal, but opposite directions.

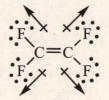

3.123 (a) Boron has four B-H bonds with two electrons in each bond, which adds up to eight valence electrons around boron.

(b)
```
       H
       |
   H – B – H
       |
       H
```

(c)

$\rangle$ 109.5°

4.1 Balance: $CO_2(g) + H_2O(l) \longrightarrow C_6H_{12}O_6(aq) + O_2(g)$
Step 1: Balance carbons with a coefficient of 6 in front of CO_2

$$6CO_2(g) + H_2O(l) \longrightarrow C_6H_{12}O_6(aq) + O_2(g)$$

Step 2: Balance hydrogen with a coefficient of 6 in front of H_2O

$$6CO_2(g) + 6H_2O(l) \longrightarrow C_6H_{12}O_6(aq) + O_2(g)$$

Step 3: Last step, balance oxygen with a coefficient of 6 in front of O_2

Balanced equation: $6CO_2(g) + 6H_2O(l) \xrightarrow{\text{photosynthesis}} C_6H_{12}O_6(aq) + 6O_2(g)$

4.2 Balance: $C_6H_{14}(l) + O_2(g) \longrightarrow CO_2(g) + H_2O(g)$
Step 1: Balance carbon with a coefficient of 6 in front of CO_2

$$C_6H_{14} + O_2 \longrightarrow 6CO_2 + H_2O$$

Step 2: Balance hydrogen with a coefficient of 7 in front of H_2O

$$C_6H_{14} + O_2 \longrightarrow 6CO_2 + 7H_2O$$

Step 3: Balance the oxygen with a coefficient of $\dfrac{19}{2}$ in front of O_2

$$C_6H_{14} + \frac{19}{2}O_2 \longrightarrow 6CO_2 + 7H_2O$$

Step 4: Although the equation is balanced, we need the coefficients to be in whole numbers, therefore we multiply each side of the equation by 2:

Balanced equation: $2C_6H_{14}(l) + 19O_2(g) \longrightarrow 12CO_2(g) + 14H_2O(g)$

4.3 Balance the equation: $K_2C_2O_4(aq) + Ca_3(AsO_4)_2(s) \longrightarrow K_3AsO_4(aq) + CaC_2O_4(s)$
Step 1: First balance the most complicated AsO_4 with a 2 in front of K_3AsO_4

$$K_2C_2O_4(aq) + Ca_3(AsO_4)_2(s) \longrightarrow 2K_3AsO_4(aq) + CaC_2O_4(s)$$

Step 2: Next, balance the potassium by placing a 3 in front of $K_2C_2O_4$

$$3K_2C_2O_4(aq) + Ca_3(AsO_4)_2(s) \longrightarrow 2K_3AsO_4(aq) + CaC_2O_4(s)$$

Step 3: Now balance C_2O_4 and Ca with a coefficient of 3 in front of CaC_2O_4
Balanced equation:

$$3K_2C_2O_4(aq) + Ca_3(AsO_4)_2(s) \longrightarrow 2K_3AsO_4(aq) + 3CaC_2O_4(s)$$

4.4 **Overall chemical reaction:** $CuCl_2(aq) + K_2S(aq) \rightarrow CuS(s) + 2KCl(aq)$
Step 1: Write an equation involving all of the chemical species participating in the chemical reaction.

$$Cu^{2+}(aq) + 2Cl^-(aq) + 2K^+(aq) + S^{2-}(aq) \rightarrow CuS(s) + 2K^+(aq) + 2Cl^-(aq)$$

Step 2: Cross out the aqueous ions that appear on both sides of the equation.

$$Cu^{2+}(aq) + \cancel{2Cl^-(aq)} + \cancel{2K^+(aq)} + S^{2-}(aq) \rightarrow CuS(s) + \cancel{2K^+(aq)} + \cancel{2Cl^-(aq)}$$

Net ionic equation: $Cu^{2+}(aq) + S^{2-}(aq) \rightarrow CuS(s)$

The net ionic equation shows the chemical species that actually undergo a chemical change. The ions that appear on both sides of the equation do not change, therefore are considered spectator ions.

4.5 (a) Ni^{2+} changed to Ni^0 by gaining 2 electrons; therefore Ni^{2+} is reduced (oxidizing agent). Cr^0 changed to Cr^{2+} by losing 2 electrons, thus Cr^0 is oxidized (reducing agent).

(b) CH_2O gains hydrogen going to CH_3OH; therefore it is reduced (oxidizing agent). H_2 "loses" hydrogen; therefore it is oxidized (reducing agent).

4.6 (a) Ibuprofen, $C_{13}H_{18}O_2$
    C 13 x 12.0 amu = 156 amu
    H 18 x 1.01 amu = 18.2 amu
    O 2 x 16.0 amu =   32.0 amu
    $C_{13}H_{18}O_2$ = 206 amu

(b) Barium Phosphate, $Ba_3(PO_4)_2$
    Ba 3 x 137 amu = 411 amu
    P   2 x 31.0 amu = 62.0 amu
    O   8 x 16.0 amu = 128 amu
    $Ba_3(PO_4)_2$ = 601 amu

4.7 $1500. \text{ g of } H_2O \left(\dfrac{1.00 \text{ mol } H_2O}{18.0 \text{ g } H_2O} \right) = 83.3 \text{ mol of } H_2O$

4.8 $2.84 \text{ mol } Na_2S \left(\dfrac{78.05 \text{ g } Na_2S}{1 \text{ mol } Na_2S} \right) = 222 \text{ g } Na_2S$

4.9 Moles of C atoms in 2.5 moles of glucose:

$$2.5 \text{ mol glucose} \left(\frac{6 \text{ mol C atoms}}{1 \text{ mol glucose}} \right) = 15 \text{ mol C atoms}$$

Moles of H atoms in 2.5 moles of glucose:

$$2.5 \text{ mol glucose} \left(\frac{12 \text{ mol C atoms}}{1 \text{ mol glucose}} \right) = 30. \text{ mol H atoms}$$

Moles of O atoms in 2.5 moles of glucose:

$$2.5 \text{ mol glucose} \left(\frac{6 \text{ mol O atoms}}{1 \text{ mol glucose}} \right) = 15 \text{ mol O atoms}$$

4.10 Moles of Cu(I) ions:

$$0.062 \text{ g CuNO}_3 \left(\frac{1 \text{ mol CuNO}_3}{125.6 \text{ g CuNO}_3} \right) \left(\frac{1 \text{ mol Cu}^+ \text{ ions}}{1 \text{ mol CuNO}_3} \right) = 4.9 \times 10^{-4} \text{ mol Cu}^+$$

4.11 Molecules of H_2O:

$$235 \text{ g H}_2\text{O} \left(\frac{1 \text{ mol H}_2\text{O}}{18.0 \text{ g H}_2\text{O}} \right) \left(\frac{6.02 \times 10^{23} \text{ molecules H}_2\text{O}}{1 \text{ mol H}_2\text{O}} \right) = 7.86 \times 10^{24} \text{ molecules}$$

4.12 (a) Balance: $Al_2O_3(s) \xrightarrow{\text{electrolysis}} Al(s) + O_2(g)$

Step 1: Balance aluminum and oxygen

$$Al_2O_3(s) \xrightarrow{\text{electrolysis}} 2Al(s) + 3/2 O_2(g)$$

Step 2: Equation is balanced, but we need to convert coefficients to whole numbers by multiplying both sides by 2

Balanced equation: $2Al_2O_3(s) \xrightarrow{\text{electrolysis}} 4Al(s) + 3O_2(g)$

(b) It requires 51 g of alumina, Al_2O_3, to produce 27 g aluminum:

$$27 \text{ g Al} \left(\frac{1 \text{ mol Al}}{27 \text{ g Al}} \right) \left(\frac{2 \text{ mol Al}_2\text{O}_3}{4 \text{ mol Al}} \right) \left(\frac{102 \text{ g Al}_2\text{O}_3}{1 \text{ mol Al}_2\text{O}_3} \right) = 51 \text{ g Al}_2\text{O}_3$$

4.13 Using the balanced equation: $CH_3OH(g) + CO(g) \longrightarrow CH_3COOH(l)$,
the molar ratio of CO required to produce CH_3COOH is 1:1; therefore 16.6 moles of CO is
required to produce 16.6 moles of CH_3COOH.

4.14 Using the balanced equation: $C_2H_4(g) + H_2O(l) \xrightarrow{\text{acid catalyst}} CH_3CH_2OH(l)$,

$$7.24 \text{ mol } C_2H_4 \left(\frac{1 \text{ mol ethanol}}{1 \text{ mol } C_2H_4} \right) \left(\frac{46.1 \text{ g ethanol}}{1 \text{ mol ethanol}} \right) = 334 \text{ g ethanol produced}$$

4.15 6.0 g carbon = 0.50 mol of carbon   2.1 g H_2 = 1.1 mol H_2

$$\text{Mass of } H_2 \text{ required} = 6.0 \text{ g C} \left(\frac{1 \text{ mol C}}{12.0 \text{ g C}} \right) \left(\frac{2 \text{ mol } H_2}{1 \text{ mol C}} \right) \left(\frac{2.0 \text{ g}}{1 \text{ mol } H_2} \right) = 2.0 \text{ g } H_2$$

$$\text{Mass of C required} = 2.1 \text{ g } H_2 \left(\frac{1 \text{ mol } H_2}{2.0 \text{ g } H_2} \right) \left(\frac{1 \text{ mol C}}{2 \text{ mol } H_2} \right) \left(\frac{12.0 \text{ g C}}{1 \text{ mol C}} \right) = 6.2 \text{ g C}$$

(a) Hydrogen is in excess and carbon is the limiting reagent.
(b) 8.0 grams of CH_4 are produced.

$$6.0 \text{ g C} \left(\frac{1 \text{ mol C}}{12.0 \text{ g C}} \right) \left(\frac{1 \text{ mol } CH_4}{1 \text{ mol C}} \right) \left(\frac{16.0 \text{ g } CH_4}{1 \text{ mol } CH_4} \right) = 8.0 \text{ g } CH_4 \text{ produced}$$

4.16 $\% \text{ yield} = \left(\dfrac{\text{Actual yield}}{\text{Theoretical yield}} \right) \times 100\%$

$\% \text{ yield} = \left(\dfrac{124.3 \text{ g aspirin isolated}}{153.7 \text{ g aspirin theoretical yield}} \right) \times 100\% = 80.87\%$

4.17 The following are the balanced equations:
(a) $HI + NaOH \rightarrow NaI + H_2O$
(b) $Ba(NO_3)_2 + H_2S \rightarrow BaS + 2HNO_3$
(c) $CH_4 + 2O_2 \rightarrow CO_2 + 2H_2O$
(d) $2C_4H_{10} \ 13O_2 \rightarrow 8CO_2 + 10H_2O$
(e) $2Fe + 3CO_2 \rightarrow Fe_2O_3 + 3CO$

4.19 $CO_2(g) + Ca(OH)_2(aq) \rightarrow CaCO_3(s) + H_2O(l)$

Chapter 4: Chemical Reactions

<u>4.21</u> $2Mg(s) + O_2(g) \rightarrow 2MgO(s)$

<u>4.23</u> $2C(s) + O_2(g) \rightarrow 2CO(g)$

<u>4.25</u> $2AsH_3(g) \xrightarrow{heat} 2As(s) + 3H_2(g)$

<u>4.27</u> $2NaCl(aq) + 2H_2O(l) \xrightarrow{electrolysis} Cl_2(g) + 2NaOH(aq) + H_2(g)$

<u>4.29</u> The following chemical reactions are balanced ionic equations:
(a) $Ag^+(aq) + Br^-(aq) \rightarrow AgBr(s)$
(b) $Cd^{2+}(aq) + S^{2-}(aq) \rightarrow CdS(s)$
(c) $2Sc^{3+}(aq) + 3SO_4^{2-}(aq) \rightarrow Sc_2(SO_4)_3(s)$
(d) $Sn^{2+}(aq) + 2Fe^{2+}(aq) \rightarrow Sn(s) + 2Fe^{3+}(aq)$ [remember, charges must also balance]
(e) $2K(s) + 2H_2O(l) \rightarrow 2K^+(aq) + 2OH^-(aq) + H_2(g)$

<u>4.31</u> (a) $Ca_3(PO_4)_2$ will form a precipitate, KCl is soluble in water.
(b) No precipitate will form (Group I chlorides and sulfates are soluble in water).
(c) $BaCO_3$ will form a precipitate, NH_4NO_3 is soluble (all nitrates are soluble in water).
(d) $Fe(OH)_2$ will form a precipitate, KCl is soluble in water.
(e) $Ba(OH)_2$ will form a precipitate, $NaNO_3$ is soluble in water.
(f) Sb_2S_3 will form a precipitate, NaCl is soluble in water.
(g) $PbSO_4$ will form a precipitate, KCl is soluble in water.

<u>4.33</u> $2Na^+(aq) + SO_3^{2-}(aq) + 2H^+(aq) + 2Cl^-(aq) \rightarrow$

$SO_2(g) + H_2O(l) + 2Na^+(aq) + 2Cl^-(aq)$

Balanced net ionic equation: $SO_3^{2-}(aq) + 2H^+(aq) \rightarrow SO_2(g) + H_2O(l)$

<u>4.35</u> (a) KCl (soluble): all Group 1 chlorides are soluble.
(b) NaOH (soluble): all sodium salts are soluble.
(c) $BaSO_4$ (insoluble): most sulfates are insoluble.
(d) Na_2SO_4 (soluble): all sodium salts are soluble.
(e) Na_2CO_3 (soluble): all sodium salts are soluble.
(f) $Fe(OH)_2$ (insoluble): most hydroxides are insoluble.

33

© 2013 Cengage Learning. All Rights Reserved. May not be scanned, copied or duplicated, or posted to a publicly accessible website, in whole or in part.

4.37 (a), (b), (c), (d), (e), (h), (i), (j), (k), (l), (m) and (n): True
  (f) False: Oxidation can be defined as the gain of oxygen atom(s) or the loss of hydrogen atom(s).
  (g) False: Reduction can be defined as the loss of oxygen atom(s) or the gain of hydrogen atom(s).

4.39 (a) C_7H_{12} is oxidized (the carbons gain oxygen going to CO_2) and O_2 is reduced.
  (b) O_2 is the oxidizing agent and C_7H_{12} is the reducing agent.

4.41 (c), (d), and (e): True
  (a) False: formula weights are expressed in atomic mass units (amu).
  (b) False: 1 amu is the mass of one atom of carbon (12) which is 1.66054×10^{-24} grams.

4.43 (a) $C_{12}H_{22}O_{11} = 342.3$ amu (b) $C_2H_5NO_2 = 75.07$ amu (c) $C_{14}H_9Cl_5 = 354.5$ amu

4.45 (a) $32 \text{ g CH}_4 \left(\dfrac{1 \text{ mole CH}_4}{16.0 \text{ g CH}_4} \right) = 2.0 \text{ mol CH}_4$

  (b) $345.6 \text{ g NO} \left(\dfrac{1 \text{ mol NO}}{30.01 \text{ g NO}} \right) = 11.52 \text{ mol NO}$

  (c) $184.4 \text{ g ClO}_2 \left(\dfrac{1 \text{ mol ClO}_2}{67.45 \text{ g ClO}_2} \right) = 2.734 \text{ mol ClO}_2$

  (d) $720. \text{ g C}_3H_8O_3 \left(\dfrac{1 \text{ mol C}_3H_8O_3}{92.1 \text{ g C}_3H_8O_3} \right) = 7.82 \text{ mol C}_3H_8O_3$

4.47 (a) $18.1 \text{ mol CH}_2O \left(\dfrac{1 \text{ mol O atoms}}{1 \text{ mol CH}_2O} \right) = 18.1 \text{ mol O atoms}$

  (b) $0.41 \text{ mol CHBr}_3 \left(\dfrac{3 \text{ mol Br atoms}}{1 \text{ mol CHBr}_3} \right) = 1.2 \text{ mol Br atoms}$

  (c) $3.5 \times 10^3 \text{ mol Al}_2(SO_4)_3 \left(\dfrac{12 \text{ O atoms}}{1 \text{ mol Al}_2(SO_4)_3} \right) = 4.2 \times 10^4 \text{ mol O atoms}$

  (d) $87 \text{ g HgO} \left(\dfrac{1 \text{ mol HgO}}{216.6 \text{ g HgO}} \right) \left(\dfrac{1 \text{ mol Hg atoms}}{1 \text{ mol HgO}} \right) = 0.40 \text{ mol Hg atoms}$

4.49 (a) 25.0 g TNT $\left(\dfrac{1 \text{ mol TNT}}{227.1 \text{ g TNT}} \right) \left(\dfrac{3 \text{ mol N atoms}}{1 \text{ mol TNT}} \right)$ = 0.330 mol N atoms

0.330 mol N atoms $\left(\dfrac{6.02 \times 10^{23} \text{ N atoms}}{1 \text{ mol N atoms}} \right)$ = 1.99 $\times$ 10^{23} N atoms

(b) 40.0 g Ethanol $\left(\dfrac{1 \text{ mol Ethanol}}{46.1 \text{ g Ethanol}} \right) \left(\dfrac{2 \text{ mol C atoms}}{1 \text{ mol Ethanol}} \right)$ = 1.74 mol C atoms

1.74 mol C atoms $\left(\dfrac{6.02 \times 10^{23} \text{ C atoms}}{1 \text{ mol C atoms}} \right)$ = 1.04 $\times$ 10^{24} C atoms

(c) 500. mg Asp $\left(\dfrac{1 \text{ g Asp}}{1000 \text{ mg Asp}} \right) \left(\dfrac{1 \text{ mol Asp}}{180.2 \text{ g Asp}} \right)$ = 2.77 $\times$ 10^{-3} mol Aspirin

2.77 $\times$ 10^{-3} mol Asp $\left(\dfrac{4 \text{ mol O atoms}}{1 \text{ mol Asp}} \right) \left(\dfrac{6.02 \times 10^{23} \text{ atoms}}{1 \text{ mol O atoms}} \right)$ = 6.67 $\times$ 10^{21} O atoms

(d) 2.40 g NaH$_2$PO$_4$ $\left(\dfrac{1 \text{ mol NaH}_2\text{PO}_4}{120.0 \text{ g NaH}_2\text{PO}_4} \right) \left(\dfrac{1 \text{ mol Na atoms}}{1 \text{ mol NaH}_2\text{PO}_4} \right)$ = 0.0200 mol Na atoms

0.0200 mol Na atoms $\left(\dfrac{6.02 \times 10^{23} \text{ Na atoms}}{1 \text{ mol Na atoms}} \right)$ = 1.20 $\times$ 10^{22} Na atoms

4.51 (a) $100. \text{ molec. CH}_2\text{O} \left(\dfrac{1 \text{ mol CH}_2\text{O}}{6.02 \times 10^{23} \text{ molec. CH}_2\text{O}} \right) = 1.66 \times 10^{-22} \text{ mol CH}_2\text{O}$

$1.66 \times 10^{-22} \text{ mol CH}_2\text{O} \left(\dfrac{30.0 \text{ g CH}_2\text{O}}{1 \text{ mol CH}_2\text{O}} \right) = 4.99 \times 10^{-21} \text{ g CH}_2\text{O}$

(b) $3000. \text{ molec. CH}_2\text{O} \left(\dfrac{1 \text{ mol CH}_2\text{O}}{6.022 \times 10^{23} \text{ molec. CH}_2\text{O}} \right) = 4.981 \times 10^{-21} \text{ mol CH}_2\text{O}$

$4.98 \times 10^{-21} \text{ moles CH}_2\text{O} \left(\dfrac{30.0 \text{ g CH}_2\text{O}}{1 \text{ mol CH}_2\text{O}} \right) = 1.496 \times 10^{-19} \text{ g CH}_2\text{O}$

(c) $5.0 \times 10^6 \text{ molec. CH}_2\text{O} \left(\dfrac{1 \text{ mol CH}_2\text{O}}{6.02 \times 10^{23} \text{ molec. CH}_2\text{O}} \right) = 8.3 \times 10^{-18} \text{ mol CH}_2\text{O}$

$8.3 \times 10^{-18} \text{ mol CH}_2\text{O} \left(\dfrac{30.0 \text{ g CH}_2\text{O}}{1 \text{ mol CH}_2\text{O}} \right) = 2.5 \times 10^{-16} \text{ g CH}_2\text{O}$

(d) $2.0 \times 10^{24} \text{ molec. CH}_2\text{O} \left(\dfrac{1 \text{ mol CH}_2\text{O}}{6.02 \times 10^{23} \text{ molec. CH}_2\text{O}} \right) = 3.3 \text{ moles CH}_2\text{O}$

$3.3 \text{ mol CH}_2\text{O} \left(\dfrac{30.0 \text{ g CH}_2\text{O}}{1 \text{ mol CH}_2\text{O}} \right) = 1.0 \times 10^2 \text{ g CH}_2\text{O}$

4.53 $3.9 \text{ mg chol.} \left(\dfrac{1 \text{ g chol.}}{1000 \text{ mg chol.}} \right) \left(\dfrac{1 \text{ mol chol.}}{386.7 \text{ g chol.}} \right) = 1.0 \times 10^{-5} \text{ mol cholesterol}$

$1.0 \times 10^{-5} \text{ mol chol.} \left(\dfrac{6.02 \times 10^{23} \text{ molec. chol.}}{1 \text{ mol chol.}} \right) = 6.1 \times 10^{18} \text{ molecules of cholesterol}$

4.55 Using the balanced equation: $2N_2(g) + 3O_2(g) \rightarrow 2N_2O_3(g)$

(a) $1 \text{ mol O}_2 \left(\dfrac{2 \text{ mol N}_2}{3 \text{ mol O}_2} \right) = 0.67 \text{ mol N}_2$ required

(b) $1 \text{ mol O}_2 \left(\dfrac{2 \text{ N}_2\text{O}_3}{3 \text{ mol O}_2} \right) = 0.67 \text{ mol N}_2\text{O}_3$ produced

(c) $8 \text{ mol N}_2\text{O}_3 \left(\dfrac{3 \text{ mol O}_2}{2 \text{ mol N}_2\text{O}_3} \right) = 12 \text{ mol O}_2$ required

4.57 Using the balanced equation: $CH_4(g) + 3Cl_2(l) \rightarrow CHCl_3(g) + 3HCl(g)$

$1.50 \text{ mol CHCl}_3 \left(\dfrac{3 \text{ mol Cl}_2}{1 \text{ mol CHCl}_3} \right) \left(\dfrac{70.9 \text{ g Cl}_2}{1 \text{ mol Cl}_2} \right) = 319 \text{ g Cl}_2$ needed

4.59 (a) Balanced equation: $2NaClO_2(aq) + Cl_2(g) \rightarrow 2ClO_2(g) + 2NaCl(aq)$

(b) 4.10 kg ClO_2

$5.50 \text{ kg NaClO}_2 \left(\dfrac{1000 \text{ g NaClO}_2}{1 \text{ kg NaClO}_2} \right) \left(\dfrac{1 \text{ mol NaClO}_2}{90.4 \text{ g NaClO}_2} \right) = 60.8 \text{ mol NaClO}_2$

$60.8 \text{ mol NaClO}_2 \left(\dfrac{2 \text{ mol ClO}_2}{2 \text{ mol NaClO}_2} \right) \left(\dfrac{67.5 \text{ g ClO}_2}{1 \text{ mol ClO}_2} \right) \left(\dfrac{1 \text{ kg}}{1000 \text{ g}} \right) = 4.10 \text{ kg ClO}_2$

4.61 Using the balanced equation: $6CO_2(g) + 6H_2O(l) \rightarrow C_6H_{12}O_6(aq) + 6O_2(g)$

$5.1 \text{ g glucose} \left(\dfrac{1 \text{ mol glucose}}{180. \text{ g glucose}} \right) \left(\dfrac{6 \text{ mol CO}_2}{1 \text{ mol glucose}} \right) \left(\dfrac{44.0 \text{ g CO}_2}{1 \text{ mol CO}_2} \right) = 7.5 \text{ g CO}_2$

4.63 Using the balanced equation in problem #4.62

$0.58 \text{ g Fe}_2\text{O}_3 \left(\dfrac{1 \text{ mol Fe}_2\text{O}_3}{159.7 \text{ g Fe}_2\text{O}_3} \right) \left(\dfrac{6 \text{ mol C}}{2 \text{ mol Fe}_2\text{O}_3} \right) \left(\dfrac{12.0 \text{ g C}}{1 \text{ mol C}} \right) = 0.13 \text{ g C}$ needed

4.65 50.0 $\cancel{\text{g asp actual yield}}$ $\left(\dfrac{100 \ \cancel{\text{g asp}}}{75 \ \cancel{\text{g asp actual yield}}} \right)\left(\dfrac{1 \text{ mol asp}}{180.2 \ \cancel{\text{g asp}}} \right) = 0.370 \text{ mol asp}$

0.370 $\cancel{\text{mol asp}}$ $\left(\dfrac{1 \ \cancel{\text{mol SA}}}{1 \ \cancel{\text{mol asp}}} \right)\left(\dfrac{138.1 \text{ g SA}}{1 \ \cancel{\text{mol SA}}} \right) = 51 \text{ g SA}$

You will need to use 51 g salicylic acid to get 50.0 g of aspirin after a 75% yield.

4.67 Using the balanced equation: $CH_3CH_3(g) + Cl_2(g) \rightarrow CH_3CH_2Cl(l) + HCl(g)$

Theoretical yield of ethyl chloride:

5.6 $\cancel{\text{g Ethane}}$ $\left(\dfrac{1 \ \cancel{\text{mol Ethane}}}{30.1 \ \cancel{\text{g Ethane}}} \right)\left(\dfrac{1 \ \cancel{\text{mol CH}_3\text{CH}_2\text{Cl}}}{1 \ \cancel{\text{mol Ethane}}} \right)\left(\dfrac{64.5 \text{ g CH}_3\text{CH}_2\text{Cl}}{1 \ \cancel{\text{mol CH}_3\text{CH}_2\text{Cl}}} \right) = 12 \text{ g}$

Actual yield of $CH_3CH_2Cl = 8.2$ g $\% \text{ yield} = \dfrac{\text{Actual yield}}{\text{Theoretical yield}} \times 100\%$

$\% \text{ Yield of } CH_3CH_2Cl = \dfrac{8.2 \text{ g CH}_3\text{CH}_2\text{Cl}}{12 \text{ g CH}_3\text{CH}_2\text{Cl}} \times 100\% = 68\%$

4.69 (a), (c), (d), (e), and (f): True

(b) False: exothermic reactions give off heat. Endothermic reactions require heat.

4.71 (a) Endothermic (22.0 kcal appears as a reactant)

(b) Exothermic (124 kcal appears as a product)

(c) Exothermic (94.0 kcal appears as a product)

(d) Endothermic (9.80 kcal appears as a reactant)

(e) Exothermic (531 kcal appears as a product)

4.73 0.37 $\cancel{\text{mol Acetone}}$ $\left(\dfrac{853.6 \text{ kcal}}{2 \ \cancel{\text{mol Acetone}}} \right) = 1.6 \times 10^2 \text{ kcal heat evolved}$

4.75 Ethanol has a greater heat of combustion per gram (7.09 kcal/g) than glucose (3.72 kcal/g).

Glucose heat of combustion (kcal/g) = $\dfrac{670. \text{ kcal/} \cancel{\text{mol}}}{180.2 \text{ g/} \cancel{\text{mol}}} = 3.72 \text{ kcal/g}$

Ethanol heat of combustion (kcal/g) = $\dfrac{327 \text{ kcal/} \cancel{\text{mol}}}{46.1 \text{ g/} \cancel{\text{mol}}} = 7.09 \text{ kcal/g}$

4.77 Using the balanced equation:

$$156.0 \text{ kcal} \left(\frac{1 \text{ mol Fe}_2O_3}{196.5 \text{ kcal}} \right) \left(\frac{2 \text{ mol Fe}}{1 \text{ mol Fe}_2O_3} \right) \left(\frac{55.85 \text{ g Fe}}{1 \text{ mol Fe}} \right) = 88.68 \text{ g Fe metal produced}$$

4.79 Hydroxyapatite is composed of calcium ions, phosphate ions, and hydroxide ions.

4.81 The unbalanced equation: $\quad$ Li $+$ I$_2$ $\quad\rightarrow\quad$ LiI

$\qquad$ oxidation: $\qquad\qquad$ $2\text{Li} \rightarrow 2\text{Li}^+ + 2e^-$

$\qquad$ reduction: $\qquad\qquad$ $\text{I}_2 + 2e^- \rightarrow 2\text{I}^-$

$\qquad$ overall reaction: $\qquad$ $2\text{Li} + \text{I}_2 \rightarrow 2\text{LiI}$

4.83 Cu^+ is oxidized (loses an electron) to Cu^{2+}. A species that is oxidized during the course of a reaction is a reducing agent, therefore Cu^+ is a reducing agent.

4.85 The combustion of organic compounds containing carbon, hydrogen, and oxygen produces carbon dioxide and water. $\qquad$ $\text{C}_5\text{H}_{12}\text{O} + \text{O}_2 \rightarrow \text{CO}_2 + \text{H}_2\text{O}$ unbalanced

1. First balance the compounds that only show up in one place on both sides of the equation (carbons): $\qquad$ $\text{C}_5\text{H}_{12}\text{O} + \text{O}_2 \rightarrow 5\text{CO}_2 + \text{H}_2\text{O}$

2. Next, balance the hydrogens:

$$\text{C}_5\text{H}_{12}\text{O} + \text{O}_2 \rightarrow 5\text{CO}_2 + 6\text{H}_2\text{O} \quad \text{unbalanced}$$

3. Now balance the oxygens: $\quad$ $\text{C}_5\text{H}_{12}\text{O} + \dfrac{15}{2}\text{O}_2 \rightarrow 5\text{CO}_2 + 6\text{H}_2\text{O}$

4. The equation is now balanced, but fractional coefficients can be awkward. By multiplying each side of the equation by 2, the coefficients will now be whole numbers. $\qquad$ $2\text{C}_5\text{H}_{12}\text{O} + 15\text{O}_2 \rightarrow 10\text{CO}_2 + 12\text{H}_2\text{O}$

4.87 $488 \text{ mg aspirin} \left(\dfrac{1 \text{ g aspirin}}{1000 \text{ mg aspirin}} \right) \left(\dfrac{1 \text{ mol aspirin}}{180.2 \text{ g aspirin}} \right) = 2.71 \times 10^{-3} \text{ mol aspirin}$

Chapter 4: Chemical Reactions

4.89 From the balanced equation: $N_2(g) + 3H_2(g) \rightarrow 2NH_3(g)$

7.0 kg of N_2 = 7.0 x 10^3 g of N_2 and 11.0 kg of H_2 = 11.0 x 10^3 g of H_2

Mass of nitrogen required based on 11.0 x 10^3 g of H_2:

$$11.0 \times 10^3 \text{ g H}_2 \left(\frac{1 \text{ mol H}_2}{2.02 \text{ g H}_2}\right)\left(\frac{1 \text{ mol N}_2}{3 \text{ mol H}_2}\right)\left(\frac{28.0 \text{ g N}_2}{1 \text{ mol N}_2}\right) = 51.0 \times 10^3 \text{ g N}_2$$

Mass of hydrogen required based on 7.0 x 10^3 g of N_2:

$$7.0 \times 10^3 \text{ g N}_2 \left(\frac{1 \text{ mol N}_2}{28.0 \text{ g N}_2}\right)\left(\frac{3 \text{ mol H}_2}{1 \text{ mol N}_2}\right)\left(\frac{2.02 \text{ g H}_2}{1 \text{ mol H}_2}\right) = 1.5 \times 10^3 \text{ g H}_2$$

Nitrogen is the limiting reagent.

4.91 4 x 10^{10} molecules of hemoglobin in a blood cell

$$2 \times 10^{-8} \text{ g cell} \left(\frac{0.20 \text{ g hem}}{1 \text{ g cell}}\right)\left(\frac{1 \text{ mol hem}}{68,000 \text{ g hem}}\right) = 6 \times 10^{-14} \text{ mol of hemoglobin}$$

$$6 \times 10^{-14} \text{ mol heme} \left(\frac{6.02 \times 10^{23} \text{ molec.}}{1 \text{ mol heme}}\right) = 4 \times 10^{10} \text{ molecules of hemoglobin}$$

4.93 From the balanced equation: $N_2(g) + 3H_2(g) \rightarrow 2NH_3(g)$

(a) 29.7 kg of N_2 = 29.7 x 10^3 g of N_2 and 3.31 kg of H_2 = 3.31 x 10^3 g of H_2

Mass of nitrogen required based on 3.31 x 10^3 g of H_2:

$$3.31 \times 10^3 \text{ g H}_2 \left(\frac{1 \text{ mol H}_2}{2.02 \text{ g H}_2}\right)\left(\frac{1 \text{ mol N}_2}{3 \text{ mol H}_2}\right)\left(\frac{28.0 \text{ g N}_2}{1 \text{ mol N}_2}\right) = 15.3 \times 10^3 \text{ g N}_2$$

Mass of hydrogen required 29.7 x 10^3 g of N_2:

$$29.7 \times 10^3 \text{ g N}_2 \left(\frac{1 \text{ mol N}_2}{28.0 \text{ g N}_2}\right)\left(\frac{3 \text{ mol H}_2}{1 \text{ mol N}_2}\right)\left(\frac{2.02 \text{ g H}_2}{1 \text{ mol H}_2}\right) = 6.41 \times 10^3 \text{ g H}_2$$

If 29.7 kg of N_2 were used in this reaction, 6.41 kg of H_2 would be required for the reaction to go to completion, which is less H_2 than the 3.31 kg that was used. Therefore, H_2 is the limiting reagent and N_2 is used in excess.

(b) If 29.7 kg of H_2 is used in the reaction, 15.3 kg of N_2 will be required. Subtract 15.3 kg of N_2 from the original 29.7 kg of N_2, 29.7 kg – 15.3 kg = 14.4 kg N_2 left over.

(c) From the above balanced equation and H_2 as the limiting reagent, the amount of NH_3 produced can be calculated as:

$$3.31 \times 10^3 \text{ g H}_2 \left(\frac{1 \text{ mol H}_2}{2.02 \text{ g H}_2}\right)\left(\frac{2 \text{ mol NH}_3}{3 \text{ mol H}_2}\right)\left(\frac{17.0 \text{ g NH}_3}{1 \text{ mol NH}_3}\right) = 18.6 \times 10^3 \text{ g NH}_3$$

4.95 (a) moles of furan = 441 mg furan $\left(\dfrac{1\ g}{1000\ mg}\right)\left(\dfrac{1\ mol\ fuan}{68.07\ g\ furan}\right)$ = 6.48 x 10^{-3} mol furan

(b) moles of carbon atoms in furan sample=

0.060 L furan $\left(\dfrac{1000\ mL}{1\ L}\right)\left(\dfrac{0.936\ g}{1\ mL}\right)\left(\dfrac{1\ mol\ furan}{68.07\ g\ furan}\right)\left(\dfrac{4\ mol\ C\ atoms}{1\ mol\ furan}\right)$ = 3.3 moles C

number of C atoms = 3.3 mol C atoms $\left(\dfrac{6.02\ x\ 10^{23}\ C\ atoms}{1\ mol\ C\ atoms}\right)$ = 2.0 x 10^{24} C atoms

(c) 9.86 x 10^{25} molec furan $\left(\dfrac{1\ mol}{6.02\ x\ 10^{23}\ molec}\right)\left(\dfrac{68.07\ g\ furan}{1\ mol\ furan}\right)$ = 1.11 x 10^{4} g furan

4.97 First, determine the number of moles of Cl_2 at the beginning of the reaction sequence:

966 kg Cl_2 $\left(\dfrac{1000\ g\ Cl_2}{1\ kg\ Cl_2}\right)\left(\dfrac{1\ mol\ Cl_2}{70.91\ g\ Cl_2}\right)$ = 1.36 x 10^{4} mol Cl_2

After the first step, the amount of KClO produced after 92.1% yield is calculated:

1.36 x 10^{4} mol Cl_2 $\left(\dfrac{1\ mol\ KClO}{1\ mol\ Cl_2}\right)\left(\dfrac{0.921\ mol\ KClO}{1\ mol\ KClO}\right)$ = 1.25 x 10^{4} mol KClO

After the second step, the amount of KClO produced after 86.7% yield is calculated:

1.25 x 10^{4} mol KClO $\left(\dfrac{1\ mol\ KClO_3}{3\ mol\ KClO}\right)\left(\dfrac{0.867\ mol\ KClO_3}{1\ mol\ KClO_3}\right)$ = 3.61 x 10^{3} mol $KClO_3$

After the third step, the amount of $KClO_4$ produced after 75.3% yield is calculated:

3.61 x 10^{3} mol $KClO_3$ $\left(\dfrac{3\ mol\ KClO_4}{4\ mol\ KClO_3}\right)\left(\dfrac{0.753\ mol\ KClO_4}{1\ mol\ KClO_4}\right)$ = 2.04 x 10^{3} mol $KClO_4$

At the end of the sequence, the mass of $KClO_4$ produced is calculated:

2.04 x 10^{3} mol $KClO_4$ $\left(\dfrac{138.55\ g}{1\ mol\ KClO_4}\right)$ = 2.83 x 10^{5} g $KClO_4$

4.99 (a) The following equations are balanced:

$$C_{16}H_{32}O_2(s) + 23O_2(g) \rightarrow 16CO_2(g) + 16H_2O(l) + 2385 \text{ kcal/mol}$$

$$C_6H_{12}O_6(s) + 6O_2(g) \rightarrow 6CO_2(g) + 6H_2O(l) + 670 \text{ kcal/mol}$$

(b) Heat of combustion palmitic acid (kcal/g) = $\dfrac{2385 \text{ kcal/mol}}{256.4 \text{ g/mol}}$ = 9.302 kcal/g

Heat of combustion glucose (kcal/g) = $\dfrac{670. \text{ kcal/mol}}{180.2 \text{ g/mol}}$ = 3.72 kcal/g

(c) Palmitic acid has a greater heat of combustion per mole.

(d) Palmitic acid also has greater heat of combustion per gram of material

Chapter 5: Gases, Liquids, and Solids

5.1 $P_2 = \dfrac{P_1 V_1}{V_2} = \dfrac{(0.70 \text{ atm})(3.8 \text{ L})}{6.5 \text{ L}} = 0.41 \text{ atm}$

5.2 $P_1 = \dfrac{P_2 T_1}{T_2} = \dfrac{(20.3 \text{ atm})(393 \text{ K})}{485 \text{ K}} = 16.4 \text{ atm}$

5.3 $P_2 = \dfrac{P_1 V_1 T_2}{T_1 V_2} = \dfrac{(0.92 \text{ atm})(20.5 \text{ L})(285 \text{ K})}{(296 \text{ K})(340.6 \text{ L})} = 0.053 \text{ atm}$

5.4 Ideal Gas Law: $PV = nRT$

$P = \dfrac{nRT}{V} = \dfrac{(2.00 \text{ mol})(0.0821 \text{ L} \cdot \text{atm} \cdot \text{mol}^{-1} \cdot \text{K}^{-1})(295 \text{ K})}{10.0 \text{ L}} = 4.84 \text{ atm}$

5.5 Ideal Gas Law: $PV = nRT$

$n = \dfrac{PV}{RT} = \dfrac{(1.05 \text{ atm})(10.0 \text{ L})}{(0.0821 \text{ L} \cdot \text{atm} \cdot \text{mol}^{-1} \cdot \text{K}^{-1})(303 \text{ K})} = 0.422 \text{ mol Ne}$

5.6 Ideal Gas Law: $PV = nRT$

$n = \dfrac{\text{mass}}{(\text{MW})} = \dfrac{PV}{RT}$

$\text{mass} = \dfrac{PV(\text{MW})}{RT} = \dfrac{(2.00 \text{ atm})(30.5 \text{ L})(4.003 \text{ g} \cdot \text{mol}^{-1})}{(0.0821 \text{ L} \cdot \text{atm} \cdot \text{mol}^{-1} \cdot \text{K}^{-1})(300. \text{ K})} = 9.91 \text{ g He}$

5.7 Dalton's Law of Partial Pressures:

Total pressure $(P_T) = P_{N_2} + P_{H_2O}$

$P_{H_2O} = P_T - P_{N_2} = 2.015 \text{ atm} - 1.908 \text{ atm} = 0.107 \text{ atm of } H_2O \text{ vapor}$

5.8 (a) Yes, there can be a hydrogen bond between water and methanol because the hydrogen atom in both molecules is bonded to an electronegative oxygen atom. Water can act as hydrogen bond acceptor with methanol as the hydrogen bond donor **(1)** and methanol can serve as the hydrogen bond acceptor with water as the hydrogen bond donor **(2)**.

$$H-\overset{\cdot\cdot}{\underset{H}{O}}: \text{'''''} H-\overset{\cdot\cdot}{\underset{CH_3}{O}}: \qquad H-\overset{\cdot\cdot}{\underset{CH_3}{O}}: \text{'''''} H-\overset{\cdot\cdot}{\underset{H}{O}}:$$

(1)   **(2)**

(b) No, C-H bonds cannot form hydrogen bonds.

5.9 Heat of vaporization of water = 540. cal/g

$$45.0 \text{ kcal} \left(\frac{1000 \text{ cal}}{1 \text{ kcal}} \right) \left(\frac{1 \text{ g H}_2\text{O}}{540. \text{ cal}} \right) = 83.3 \text{ g H}_2\text{O vaporized}$$

5.10 The heat required to heat 1.0 g iron to melting = $\underline{2.3 \times 10^2 \text{ cal}}$.

$$\text{Heat (up to melting)} = \left(0.11 \text{ cal} \cdot \text{g}^{-1} \cdot {}^\circ\text{C}^{-1} \right) \left(1.0 \text{ g} \right) \left(1530^\circ\text{C} - 25^\circ\text{C} \right) = 1.7 \times 10^2 \text{ cal}$$

$$\text{Heat @ melting} = (1.0 \text{ g})(63.7 \text{ cal} \cdot \text{g}^{-1}) = 6.4 \times 10^1 \text{ cal}$$

$$\text{Total heat needed during the process} = 1.7 \times 10^2 \text{ cal} + 6.4 \times 10^1 \text{ cal} = 2.3 \times 10^2 \text{ cal}$$

5.11 According to the phase diagram of water (Figure 5.20), the vapor will undergo reverse sublimation and form solid ice.

5.12 Unit conversion:

$$29.5 \text{ in Hg} \left(\frac{1 \text{ atm}}{29.92 \text{ in Hg}} \right) = 0.986 \text{ atm}$$

5.13 Kinetic molecular theory explains that as the volume of a gas decreases, the concentration of gas molecules per unit of volume increases and the number of gas molecules colliding with the walls of the container increases. Because gas pressure results from the collisions of gas molecules with the walls of the container, as volume decreases, pressure increases.

5.15 Using the ideal gas law equation, the volume of a gas can be decreased by (1) increasing the pressure on the gas, (2) lowering the temperature (cooling) of the gas, and (3) decreasing the number of moles of the gas.

5.17 Boyle's Law:

At constant temperature, $\left(\dfrac{P_1 V_1}{\cancel{T_1}}\right) = \left(\dfrac{P_2 V_2}{\cancel{T_2}}\right)$ reduces to $P_1 V_1 = P_2 V_2$

$$V_2 = \frac{P_1 V_1}{P_2} = \frac{(1.10\ \cancel{atm})(6.20\ L)}{(0.925\ \cancel{atm})} = 7.37\ L$$

5.19 Boyle's Law:

At constant temperature, $\left(\dfrac{P_1 V_1}{\cancel{T_1}}\right) = \left(\dfrac{P_2 V_2}{\cancel{T_2}}\right)$ reduces to $P_1 V_1 = P_2 V_2$

$$P_2 = \frac{P_1 V_1}{V_2} = \frac{(1.0\ atm)(20.0\ \cancel{mL})}{10.0\ \cancel{mL}} = 2.0\ atm\ of\ CO_2\ gas$$

5.21 Charles's Law:

At constant presure, $\left(\dfrac{\cancel{P_1} V_1}{T_1}\right) = \left(\dfrac{\cancel{P_2} V_2}{T_2}\right)$ reduces to $\dfrac{V_1}{T_1} = \dfrac{V_2}{T_2}$

$$T_2 = \frac{T_1 V_2}{V_1} = \frac{(283\ K)(50.0\ \cancel{L})}{23.0\ \cancel{L}} = 615\ K\ (342\ ^{\circ}C)$$

5.23 Charles's Law:

At constant presure, $\left(\dfrac{\cancel{P_1} V_1}{T_1}\right) = \left(\dfrac{\cancel{P_2} V_2}{T_2}\right)$ reduces to $\dfrac{V_1}{T_1} = \dfrac{V_2}{T_2}$

$$V_2 = \frac{V_1 T_2}{T_1} = \frac{(5.2\ L)(363\ \cancel{K})}{303\ \cancel{K}} = 6.2\ L\ of\ SO_2\ gas\ upon\ heating$$

5.25 The pressure read by the manometer is the difference between the gas in the bulb and the atmospheric pressure: 833 mm Hg – 760 mm Hg = 73 mm Hg.

5.27 Gay-Lussac's Law:

At constant volume, $\left(\dfrac{P_1 \cancel{V_1}}{T_1}\right) = \left(\dfrac{P_2 \cancel{V_2}}{T_2}\right)$ reduces to $\dfrac{P_1}{T_1} = \dfrac{P_2}{T_2}$

$$P_2 = \frac{P_1 T_2}{T_1} = \frac{(2.3\ atm)(310\ \cancel{K})}{273\ \cancel{K}} = 2.6\ atm\ of\ Halothane$$

Chapter 5: Gases, Liquids, and Solids

5.29 Complete this table: The $\frac{P_1V_1}{T_1} = \frac{P_2V_2}{T_2}$ equation applies here.

V₁	T₁	P₁	V₂	T₂	P₂
6.35 L	10°C	0.75 atm	**4.6 L**	0°C	1.0 atm
75.6 L	0°C	1.0 atm	**88 L**	35°C	735 torr
1.06 L	75°C	0.55 atm	3.2 L	0°C	**0.14 atm**

5.31 $V_2 = \frac{P_1V_1T_2}{T_1P_2} = \frac{(760 \text{ mm Hg})(1 \times 10^6 \text{ L})(240 \text{ K})}{(273 \text{ K})(243 \text{ mm Hg})} = 3 \times 10^6 \text{ L}$

5.33 $T_2 = \frac{P_2V_2T_1}{P_1V_1} = \frac{(0.500 \text{ atm})(300. \text{ K})(2 V_1)}{(1.00 \text{ atm})(V_1)} = 300. \text{ K}$

5.35 $P_2 = \frac{P_1V_1T_2}{T_1V_2} = \frac{(26.4 \text{ mL})(2.50 \text{ atm})(283 \text{ K})}{(275.5 \text{ K})(36.2 \text{ mL})} = 1.87 \text{ atm}$

5.37 (a) $n = \frac{PV}{RT} = \frac{(1.33 \text{ atm})(50.3 \text{ L})}{(0.0821 \text{ L·atm·mol}^{-1}\text{·K}^{-1})(350. \text{ K})} = 2.33 \text{ mol}$

(b) No, the only information that we need to know about the gas is that it is an ideal gas.

5.39 Using the PV=nRT ideal gas law equation, the following equation is derived:

$MW = \frac{(mass)RT}{PV} = \frac{(8.00 \text{ g})(0.0821 \text{ L·atm·mol}^{-1}\text{·K}^{-1})(273 \text{ K})}{(2.00 \text{ atm})(22.4 \text{ L})} = 4.00 \text{ g/mol}$

5.41 Using the PV=nRT equation, we can derive:

$$\frac{mass}{V} = density = \frac{P(MW)}{RT}$$

(a) At constant T, equation reduces to density = (constant) x (pressure), therefore, the density increases as pressure increases.

(b) At constant P, the equation reduces to density = (constant)(1/T), therefore, density decreases with increasing T.

46

© 2013 Cengage Learning. All Rights Reserved. May not be scanned, copied or duplicated, or posted to a publicly accessible website, in whole or in part.

5.43 Using the PV = nRT equation:

(a) $n = \dfrac{PV}{RT} = \dfrac{(3.00\ \text{atm})(200.\ \text{L})}{(0.0821\ \text{L} \cdot \text{atm} \cdot \text{mol}^{-1} \cdot \text{K}^{-1})(296\ \text{K})} = 24.7\ \text{mol}\ O_2$

(b)   Mass of O_2 = $24.7\ \text{mol}\ O_2 \left(\dfrac{32.0\ \text{g}\ O_2}{1\ \text{mol}\ O_2} \right) = 790.\ \text{g}\ O_2$

5.45 $5.5\ \text{L air} \left(\dfrac{0.21\ \text{L}\ O_2}{1\ \text{L air}} \right) = 1.2\ \text{L}\ O_2$

Moles of $O_2 = n = \dfrac{PV}{RT} = \dfrac{(1.1\ \text{atm})(1.2\ \text{L})}{(0.0821\ \text{L} \cdot \text{atm} \cdot \text{mol}^{-1} \cdot \text{K}^{-1})(310.\ \text{K})} = 0.052\ \text{mol}\ O_2$

$0.052\ \text{mol}\ O_2 \left(\dfrac{6.02 \times 10^{23}\ \text{molecules}\ O_2}{1\ \text{mol}\ O_2} \right) = 3.1 \times 10^{23}\ \text{molecules}\ O_2$

5.47 1.0000 mole air = 0.7808 mol N_2 + 0.2095 mol O_2 + 0.0093 mol Ar

1.000 mole of air =

$0.7808\ \text{mol}_{N_2} \left(\dfrac{28.01\ \text{g}\ N_2}{1\ \text{mol}\ N_2} \right) + 0.2095\ \text{mol}_{O_2} \left(\dfrac{32.00\ \text{g}\ O_2}{1\ \text{mol}\ O_2} \right) + 0.0093\ \text{mol}_{Ar} \left(\dfrac{39.95\ \text{g}\ Ar}{1\ \text{mol}\ Ar} \right)$

(a)   1.000 mole of air = 28.95 g/mol

(b) $d(g/L)_{air} = \left(\dfrac{28.95\ \text{g air}}{1\ \text{mol air}} \right) \left(\dfrac{1\ \text{mol air}}{22.4\ \text{L}} \right) = 1.29\ \text{g/L}$

5.49 (a) $d_{SO_2} = \left(\dfrac{64.1 \text{ g}}{1 \text{ mol SO}_2}\right)\left(\dfrac{1 \text{ mol SO}_2}{22.4 \text{ L}}\right) = 2.86$ g/L

(b) $d_{CH_4} = \left(\dfrac{16.0 \text{ g}}{1 \text{ mol CH}_4}\right)\left(\dfrac{1 \text{ mol CH}_4}{22.4 \text{ L}}\right) = 0.714$ g/L

(c) $d_{H_2} = \left(\dfrac{2.02 \text{ g}}{1 \text{ mol H}_2}\right)\left(\dfrac{1 \text{ mol H}_2}{22.4 \text{ L}}\right) = 0.0902$ g/L

(d) $d_{He} = \left(\dfrac{4.00 \text{ g}}{1 \text{ mol H}_2}\right)\left(\dfrac{1 \text{ mol He}}{22.4 \text{ L}}\right) = 0.179$ g/L

(e) $d_{CO_2} = \left(\dfrac{44.0 \text{ g}}{1 \text{ mol CO}_2}\right)\left(\dfrac{1 \text{ mol CO}_2}{22.4 \text{ L}}\right) = 1.96$ g/L

Comparing the gas densities with that of air (determined in problem #5.47), SO_2 and CO_2 are denser than air; He, H_2 and CH_4 are less dense than air.

5.51 1.00 mL of octane (d = 0.7025 g/mL)

Mass of octane $= 1.00 \text{ mL Octane}\left(\dfrac{0.7025 \text{ g Octane}}{1 \text{ mL Octane}}\right) = 0.7025$ g Octane

$$V_{Octane} = \dfrac{nRT}{P} = \dfrac{\left(\dfrac{0.7025 \text{ g}}{114.2 \text{ g}\cdot\text{mol}^{-1}}\right)\left(0.0821 \text{ L}\cdot\text{atm}\cdot\text{mol}^{-1}\cdot\text{K}^{-1}\right)\left(373 \text{ K}\right)}{\left(725 \text{ torr}\right)\left(\dfrac{1.00 \text{ atm}}{760 \text{ torr}}\right)} = 0.197 \text{ L}$$

5.53 First, the number of moles of H_2 gas produced is calculated:

$3.50 \text{ g Na}\left(\dfrac{1 \text{ mol Na}}{22.99 \text{ g Na}}\right)\left(\dfrac{1 \text{ mol H}_2}{2 \text{ mol Na}}\right) = 7.61 \times 10^{-2}$ mol H_2 produced

Now, apply the ideal gas law equation (PV=nRT) to calculate the volume of H_2 produced at 18°C (291 K) and 0.995 atm.

$$V = \dfrac{(7.61 \times 10^{-2} \text{ mol})(0.0821 \cdot \text{atm} \cdot \text{mol}^{-1} \cdot \text{K}^{-1})(291 \text{ K})}{(0.995 \text{ atm})} = 1.83 \text{ L } H_2 \text{ produced}$$

5.55 (a) and (c): True
   (b) False: the units for partial pressure are the same as any other pressure unit is such as atm or mm Hg.
   (d) False: the total pressure of the system would be 2.00 atm.

5.57 $P_T = P_{N2} + P_{O2} + P_{CO2} + P_{H2O}$

$P_{N_2} = (0.740)(1.0 \text{ atm}) = 0.740 \text{ atm} (562.4 \text{ mm Hg})$

$P_{O_2} = (0.194)(1.0 \text{ atm}) = 0.194 \text{ atm} (147.5 \text{ mm Hg})$

$P_{H_2O} = (0.062)(1.0 \text{ atm}) = 0.062 \text{ atm} (47.1 \text{ mm Hg})$

$P_{CO_2} = (0.004)(1.0 \text{ atm}) = 0.004 \text{ atm} (3.0 \text{ mm Hg})$

$P_T = 1.0 \text{ atm} \quad (760.0 \text{ mm Hg})$

5.59 (a), (c), (e), (f), (g), (i), and (j): True
   (b) False: the average kinetic energy is proportional to temperature in Kelvin.
   (d) False: there are no attractive intermolecular forces between gas molecules.
   (h) False: ideal gases do not exist.

5.61 (c), (d), (f), (g), (h), (m), and (n): True
   (a) False: ion-ion attractive forces are the strongest.
   (b) False: covalent bond strengths vary greatly depending on the atoms bonded.
   (e) False: London dispersion forces exist between all molecules
   (i) False: even though CO_2 contains polar C=O bonds, the individual C=O dipoles act in equal, but opposite directions, resulting in a net zero dipole moment.
   (j) False molecular dipole moments depend in the individual bond dipoles along with the geometry of these bonds in the molecule. Strength of molecular dipoles is independent of molecular weight.
   (k) False a hydrogen bond is a noncovalent force of attraction between the partial positive charge on a hydrogen atoms bonded to an F, N, or O atom and the partial negative charge on a nearby F, O, or N atom.
   (l) False an -OH hydrogen bond is about 10 kcal/mol while the O-H covalent bond is between 90-120 kcal/mol.
   (o) False: the O-H bond is more polar than the N-H bond, therefore, the intermolecular hydrogen bond between molecules containing –OH groups will be stronger than those with –NH groups.

5.63 Gases behave most ideally under low pressures and high temperatures to minimize non-ideal intermolecular interactions; therefore, choice (c) best suits these conditions.

5.65 (a) CCl₄ is nonpolar: London dispersion forces
(b) CO is polar, dipole-dipole interaction.
    The most polar molecule (CO) will have the largest surface tension.

5.67 Yes, the London dispersion forces range from 0.001-0.2 kcal/mol where the lower end of dipole-dipole attractive forces can be as low as 0.1 kcal/mol.

5.69 (a), (d), (f), (g), (h), (i), (j), (m), and (o): True
(b) False: molecules in the liquid state "slide" past each other in random motion. Liquids assume the shape of their container.
(c) False: the intermolecular forces within a liquid cause surface tension. This results in the liquid assuming a minimum surface area.
(e) False: water has a high surface tension because it is a very polar molecule.
(k) False: the temperature water boils at is the temperature where its vapor pressure is equal to the external (atmospheric) pressure.
(l) False: The most important factor in determining relative boiling points is the strength of the intermolecular attractive forces.
(n) False: There is no hydrogen bonding in hydrocarbons like hexane and methane.
(p) False: The opposite is true.

5.71 (a), (b), (c), (e), (g), and (k): True
(d) False: some solids can exist in different forms, such as graphite and diamond, which are both solid forms of carbon.
(f) False: diamond is a network solid of tetrahedrally bonded carbons atoms.
(h) False: nanotubes have lengths up to 20 mm long.
(i) False: the diameter of buckyball is 0.7 nm.
(j) False: some solids decompose before their melting temperatures.

5.73 (a), (b), (d), (e), (j), and (k): True
(c) False: the temperature of an ice water mixture upon heating will not change until all of the ice has melted.
(f) False: The danger of steam burns comes from the heat of vaporization transferred to the skin upon condensation.
(g) False: for water, the heat of fusion is 80 cal/g and the heat of vaporization is 540 cal/g.
(h) False: for water, the specific heat is 1.0 cal/g °C.
(i) False: melting is an endothermic process and crystallization is an exothermic process.

5.75 $39.2 \text{ g CF}_2\text{Cl}_2 \left(\dfrac{1 \text{ mol CF}_2\text{Cl}_2}{120.9 \text{ g CF}_2\text{Cl}_2} \right) \left(\dfrac{4.71 \text{ kcal}}{1 \text{ mol CF}_2\text{Cl}_2} \right) = 1.53 \text{ kcal to vaporize the CF}_2\text{Cl}_2$

5.77 (a) ~80 mm Hg　　(b) ~130 mm Hg　　(c) ~390 mm Hg

5.79 (a) HCl < HBr < HI
Increasing size of molecule increases London dispersion forces.
(b) O_2 < HCl < H_2O_2
O_2 has only weak London dispersion forces to overcome for boiling, where HCl is a polar molecule with stronger dipole-dipole attractions to overcome for boiling. H_2O_2 has the strongest intermolecular forces (hydrogen bonding) to exceed for boiling to occur.

5.81 The difference between heating water from 0°C to 37°C and heating ice from 0°C to 37°C is the heat of fusion.
The energy required to heat ice from 0°C to 37°C:

$$100. \text{ g } H_2O \left(\frac{1.0 \text{ cal}}{\text{g}\cdot°\text{C}}\right)(37°\text{C}) + 100. \text{ g } H_2O \left(\frac{80 \text{ cal}}{\text{g}}\right) = 1.2 \times 10^4 \text{ cal (12 kcal)}$$

The energy required to heat liquid water from 0°C to 37°C:

$$100. \text{ g } H_2O \left(\frac{1.0 \text{ cal}}{\text{g}\cdot°\text{C}}\right)(37°\text{C}) = 3.7 \times 10^3 \text{ cal (3.7 kcal)}$$

5.83 Sublimation is the conversion of a solid to gas, bypassing the liquid phase.

5.85 $1.00 \text{ mL Freon-11} \left(\frac{1.49 \text{ g Freon-11}}{1 \text{ mL Freon-11}}\right)\left(\frac{1 \text{ mol Freon-11}}{137.4 \text{ g Freon-11}}\right) = 1.08 \times 10^{-2} \text{ mol Freon-11}$

$1.08 \times 10^{-2} \text{ mol Freon-11} \left(\frac{6.42 \text{ kcal}}{1 \text{ mol Freon-11}}\right) = 6.96 \times 10^{-2} \text{ kcal}$

5.87 When a person lowers their diaphragm, the volume of the chest cavity increases, thus lowering the pressure in the lungs relative to atmospheric pressure. Air at atmospheric pressure then rushes into the lungs, beginning the breathing process.

5.89 The first tapping sound one hears is the systolic pressure, which occurs when the sphygmomanometer pressure matches the blood pressure when the ventricle contracts, pushing blood into the arm.

5.91 When water freezes, it expands (water is one of the few substances that expands upon freezing) and will break the bottle when the ice expansion exceeds the volume of the bottle.

5.93 It is difficult to compress liquids and solids because their molecules or atoms are already very close together and there is very little empty space between them.

5.95 Conversion of psi to atm of an average tire pressure of 34 psi:

$$34 \text{ psi} \left(\frac{1 \text{ atm}}{14.7 \text{ psi}} \right) = 2.3 \text{ atm}$$

5.97 Aerosol cans already contain gases under high pressures. Gay-Lussac's Law predicts that the pressure inside the can will increase with increasing temperature, with the potential of the can explosively rupturing and causing injury.

5.99 $V_2 = \dfrac{P_1 V_1 T_2}{T_1 P_2} = \dfrac{\left(275 \text{ mm Hg} \right)\left(\dfrac{1 \text{ atm}}{760 \text{ mm Hg}} \right)(387 \text{ mL})(378 \text{ K})}{(348 \text{ K})(1.36 \text{ atm})} = 112 \text{ mL}$

5.101 Boiling point order:
H_2O (H-bonding) > $CHCl_3$ (dipole-dipole) > C_5H_{12} (London dispersion forces)
Water, which forms strong intermolecular hydrogen bonding requires the most energy to break, has the highest boiling point.

5.103 (a) As a gas is compressed under pressure, the molecules are forced closer together and the intermolecular forces pull the molecules together, forming a liquid.

(b) $20 \text{ lbs propane} \left(\dfrac{1 \text{ kg}}{2.205 \text{ lb of propane}} \right) = 9.1 \text{ kg}$

(c) $9.1 \text{ kg propane} \left(\dfrac{1000 \text{ g propane}}{1 \text{ kg propane}} \right)\left(\dfrac{1 \text{ mol propane}}{44.1 \text{ g propane}} \right) = 2.1 \times 10^2 \text{ mol of propane}$

(d) $210 \text{ mol propane} \left(\dfrac{22.4 \text{ L propane}}{1 \text{ mol propane}} \right) = 4.7 \times 10^3 \text{ L propane}$

5.105 $d \text{ (g/L)} = \left(\dfrac{0.00300 \text{ g}}{\text{cm}^3} \right)\left(\dfrac{1000 \text{ cm}^3}{\text{L}} \right) = 3.00 \text{ g/L}$

$MW = \dfrac{\text{mass} RT}{VP} = \dfrac{(3.00 \text{ g})(0.0821 \text{ L} \cdot \text{atm} \cdot \text{mol}^{-1} \cdot \text{K}^{-1})(373 \text{ K})}{(1.00 \text{ L})(1.00 \text{ atm})} = 91.9 \text{ g/mol}$

5.107 Use the PV = nRT equation after converting some of the units:

$$\text{Mol of } NH_3 = 60.0 \text{ g } NH_3 \left(\frac{1 \text{ mol } NH_3}{17.0 \text{ g } NH_3} \right) = 3.52 \text{ mol } NH_3$$

$$P \text{ (in atm)} = 77.2 \text{ inch Hg} \left(\frac{25.4 \text{ mm Hg}}{1 \text{ inch Hg}} \right)\left(\frac{1 \text{ atm}}{760 \text{ mm Hg}} \right) = 2.58 \text{ atm}$$

$$T = \frac{PV}{nR} = \frac{(2.58 \text{ atm})(35.1 \text{ L})}{(3.52 \text{ mol})(0.0821 \text{ L·atm·mol}^{-1}\text{·K}^{-1})} = 313 \text{ K } (40.°C)$$

5.109 The temperature of a liquid drops during evaporation because the molecules with higher kinetic energy leave the liquid as a gas, decreasing the average kinetic energy of the liquid. The temperature of the liquid is directly proportional to its average kinetic energy, therefore, the temperature decreases as the average kinetic energy decreases.

5.111 (a) Pressure on body (from water) = P_{water} = 100 ft $\left(\frac{1 \text{ atm}}{33 \text{ ft}} \right)$ = 3.0 atm

P_{total} = Pressure on body (at depth of 100 ft) = P_{atm} + P_{water} = 4.0 atm

(b) At 1.00 atm, P_{N2} = 593 mm Hg (0.780 atm) and thus makes up 78.0% of the gas mixture, which does not change at the depth of 100 feet. At a depth of 100 feet, the total pressure on the lungs, which is equalized by pressure of air delivered by the SCUBA tank, is 4.0 atm.

$$P_{N_2} \text{ (at 100 ft)} = 4.0 \text{ atm total pressure} \left(\frac{0.780 \text{ atm } P_{N_2}}{1 \text{ atm total pressure}} \right) = 3.1 \text{ atm } P_{N_2}$$

(c) At 2 atm, P_{O2} = 158 mm Hg (0.208 atm) and thus makes up 20.8% of the gas mixture at 2 atm, which does not change at the depth of 100 feet. At a depth of 100 feet, the total pressure on the lungs, which is equalized by pressure of air delivered by the SCUBA tank, is 4.0 atm.

$$P_{O_2} \text{ (at 100 ft)} = 4.0 \text{ atm total pressure} \left(\frac{0.208 \text{ atm } P_{O_2}}{1 \text{ atm total pressure}} \right) = 0.83 \text{ atm } P_{O_2}$$

(d) As a diver ascends from 100 ft, the external pressure on the lungs decreases and therefore the volume of gasses in the lungs increases. If the diver does not exhale vigorously during a rapid ascent, the diver's lungs could over-inflate due to expanding gasses in the lungs, causing injury.

5.113 First, determine the number of moles of each reactant at the reaction conditions using the ideal gas law equation:

$$n_{NH3} = \frac{PV}{RT} = \frac{(1.02 \text{ atm})(4.21 \text{ L})}{(0.0821 \text{ L}\cdot\text{atm}\cdot\text{mol}^{-1}\cdot\text{K}^{-1})(300. \text{ K})} = 0.174 \text{ mol NH}_3$$

$$n_{HCl} = \frac{PV}{RT} = \frac{(0.998 \text{ atm})(5.35 \text{ L})}{(0.0821 \text{ L}\cdot\text{atm}\cdot\text{mol}^{-1}\cdot\text{K}^{-1})(299 \text{ K})} = 0.218 \text{ mol HCl}$$

The reactants combine in a 1:1 molar ratio to form NH_4Cl product. The limiting reagent is ammonia.

$$0.174 \text{ mol NH}_3 \left(\frac{1 \text{ mol NH}_4Cl}{1 \text{ mol NH}_3}\right)\left(\frac{53.49 \text{ g NH}_4Cl}{1 \text{ mol NH}_4Cl}\right) = 9.31 \text{ g NH}_4Cl \text{ formed}$$

5.115 First, balance the equation to: $NH_4NO_2 \rightarrow N_2 + 2H_2O$

Next, determine the number of moles of product at the reaction conditions using the unit conversions, Dalton's law of partial pressures, and the ideal gas law equation:

T(K) = 26 + 273 = 299 K   511 mL = 0.511 L

Dalton's law of partial pressures: $P_T = P_{N2} + P_{H2O}$

If the water vapor component is removed, the pressure resulting is the pressure from the N_2.

P_{N2} = 745 torr - 25.2 torr = 720. torr

$$P_{N2} = 720. \text{ torr} \left(\frac{1.00 \text{ atm}}{760. \text{ torr}}\right) = 0.947 \text{ atm}$$

$$n_{N2} = \frac{PV}{RT} = \frac{(0.947 \text{ atm})(0.511 \text{ L})}{(0.0821 \text{ L}\cdot\text{atm}\cdot\text{mol}^{-1}\cdot\text{K}^{-1})(299 \text{ K})} = 1.97 \times 10^{-2} \text{ mol N}_2$$

Now, from the balanced equation, using the amount of dry nitrogen gas produced, the starting amount of NH_4NO_2 can be determined:

$$1.97 \times 10^{-2} \text{ mol N}_2 \left(\frac{1 \text{ mol NH}_4NO_2}{1 \text{ mol N}_2}\right)\left(\frac{64.044 \text{ g NH}_4NO_2}{1 \text{ mol NH}_4NO_2}\right) = 1.26 \text{ g NH}_4NO_2$$

5.117   heat lost by cooling the gas from 120.°C to 100.°C:

$$\text{heat} = \text{mass} \times SH \times \Delta T = 5.75 \ \cancel{g}\left(0.48 \ \text{cal} \cdot \cancel{g^{-1}} \cdot \cancel{{}^{\circ}C^{-1}}\right)\left(20.^{\circ}\cancel{C}\right) = 55.2 \ \text{cal}$$

heat lost by condensation (phase change) at 100.°C involves the heat of vaporization:

$$\text{heat} = \text{mass} \times \text{heat of vaporiation} = 5.75 \ \cancel{g} \times 540. \ \text{cal} \cdot \cancel{g^{-1}} = 3105 \ \text{cal}$$

heat lost by cooling the liquid from 100.°C to 0.°C:

$$\text{heat} = \text{mass} \times SH \times \Delta T = 5.75 \ \cancel{g}\left(1.00 \ \text{cal} \cdot \cancel{g^{-1}} \cdot \cancel{{}^{\circ}C^{-1}}\right)\left(100.^{\circ}\cancel{C}\right) = 575 \ \text{cal}$$

heat lost by freezing (phase change) at 0°C involves the heat of fusion:

$$\text{heat} = \text{mass} \times \text{heat of fusion} = 5.75 \ \cancel{g} \times 80. \ \text{cal} \cdot \cancel{g^{-1}} = 460 \ \text{cal}$$

heat lost by cooling the solid from 0°C to -20.°C:

$$\text{heat} = \text{mass} \times SH \times \Delta T = 5.75 \ \cancel{g}\left(0.48 \ \text{cal} \cdot \cancel{g^{-1}} \cdot \cancel{{}^{\circ}C^{-1}}\right)\left(20.^{\circ}\cancel{C}\right) = 55.2 \ \text{cal}$$

add up the heat lost from each step: 4.25×10^3 cal of heat lost total

Chapter 6: Solutions and Colloids

6.1 4.4% w/v KBr solution = 4.4 g KBr in 100 mL of solution

$$250 \text{ mL solution}\left(\frac{4.4 \text{ g KBr}}{100 \text{ mL solution}}\right) = 11 \text{ g KBr}$$

Add enough water to 11 g KBr to make 250 mL of solution

6.2 % (w/v) = (mass of the solute/volume of solution) x 100%

$$\% \text{ (w/v)} = \left(\frac{6.7 \text{ g LiI}}{400. \text{ mL solution}}\right) \times 100\% = 1.7\%$$

6.3 First, calculate the number of moles and mass of KCl that are needed:
Moles = M x V.

$$\text{Moles of KCl} = \left(\frac{1.06 \text{ mol KCl}}{1 \text{ L sol}}\right)(2.0 \text{ L sol}) = 2.1 \text{ mol KCl}$$

$$\text{Mass of KCl} = 2.1 \text{ mol KCl}\left(\frac{74.6 \text{ g KCl}}{1 \text{ mol KCl}}\right) = 1.6 \times 10^2 \text{ g KCl}$$

Place 1.6×10^2 g of KCl into a 2.0-L volumetric flask, add some water, swirl until the solid has dissolved, and then fill the flask with water to the 2.0-L mark.

6.4 M = moles of solute/L of solution:
Because molarity units are in moles of solute/L of solution, grams of KSCN must be converted to moles of KSCN and mL of solution to L of solution:

$$M = \frac{0.440 \text{ g KSCN}}{340. \text{ mL sol}}\left(\frac{1 \text{ mol KSCN}}{97.2 \text{ g KSCN}}\right)\left(\frac{1000 \text{ mL sol}}{1 \text{ L sol}}\right) = 0.0133 \, M \text{ KSCN sol}$$

6.5 First, convert grams of glucose into moles of glucose, then convert moles of glucose into mL of solution:

$$\text{Moles of glucose} = 10.0 \text{ g glucose}\left(\frac{1 \text{ mol glucose}}{180. \text{ g glucose}}\right) = 0.0556 \text{ mol glucose}$$

$$0.0556 \text{ mol glucose}\left(\frac{1 \text{ L sol}}{0.300 \text{ mol glucose}}\right)\left(\frac{1000 \text{ mL sol}}{1 \text{ L sol}}\right) = 185 \text{ mL glucose sol}$$

6.6 First, 100. gallons must be converted to liters:

$$100.\ \text{gal}\left(\frac{3.785\ \text{L}}{1\ \text{gal}}\right) = 378.5\ \text{L} \quad \text{then, calculate grams of NaHSO}_3 \text{ in a 0.010 } M \text{ solution}$$

$$378.5\ \cancel{\text{L sol}}\left(\frac{0.010\ \text{mol NaHSO}_3}{1\ \cancel{\text{L sol}}}\right)\left(\frac{104\ \text{g NaHSO}_3}{1\ \cancel{\text{mol NaHSO}_3}}\right) = 3.9 \times 10^2 \text{ g NaHSO}_3 \text{ added}$$

6.7 Use the $M_1V_1 = M_2V_2$ equation:

$$V_1 = \frac{\left(0.600\ \cancel{M\ \text{HCl}}\right)\left(300.\ \text{mL sol HCl}\right)}{\left(12.0\ \cancel{M\ \text{HCl}}\right)} = 15.0\ \text{mL}$$

Place 15.0 mL of a 12.0 M HCl solution into a 300-mL volumetric flask, add some water, swirl until completely mixed, and then fill the flask with water to the 300-mL mark.

6.8 Use the $\%_1V_1 = \%_2V_2$ equation:

$$V_1 = \frac{\left(0.10\ \cancel{\%\ \text{KOH}}\right)\left(20.0\ \text{mL sol KOH}\right)}{\left(15\ \cancel{\%\ \text{KOH}}\right)} = 0.13\ \text{mL of 15\% KOH}$$

Place 0.13 mL of a 15% KOH solution into a 20-mL volumetric flask, add some water, swirl until completely dissolved, and then fill the flask with water to the 20-mL mark.

6.9 First calculate the mass of Na^+ ion in the 560. g of NaHSO_4. Then determine the Na^+ ion concentration in the solution using ppm.

$$560.\ \cancel{\text{g NaHSO}_4}\left(\frac{1\ \cancel{\text{mol NaHSO}_4}}{120.\ \cancel{\text{g NaHSO}_4}}\right)\left(\frac{1\ \cancel{\text{mol Na}^+}}{1\ \cancel{\text{mol NaHSO}_4}}\right)\left(\frac{23.0\ \text{g Na}^+}{1\ \cancel{\text{mol Na}^+}}\right) = 107\ \text{g Na}^+$$

$$[\text{Na}^+] = \frac{107\ \text{g Na}^+}{4.5 \times 10^5\ \cancel{\text{L sol}}}\left(\frac{1\ \cancel{\text{L sol}}}{1\ \cancel{\text{kg sol}}}\right)\left(\frac{1\ \cancel{\text{kg sol}}}{1000\ \text{g sol}}\right) \times 10^6\ \text{ppm} = 0.24\ \text{ppm Na}^+$$

6.10 $\Delta T = (1.86°\text{C/mol})(\text{mole of particles in solution per 1000. g of water})$:

$$\text{Moles of CH}_3\text{OH} = 215\ \cancel{\text{g CH}_3\text{OH}}\left(\frac{1\ \text{mol CH}_3\text{OH}}{32.0\ \cancel{\text{g CH}_3\text{OH}}}\right) = 6.72\ \text{mol CH}_3\text{OH}$$

$\Delta T = (1.86°\text{C/mol})(6.72\ \text{mol}) = 12.5°\text{C}$ freezing point will be lowered by 12.5°C

New freezing point$_{(\text{H}_2\text{O})}$ = 0°C - 12.5°C = -12.5°C

6.11 Compare the concentration of moles of ions or molecules in each solution. The solution with the highest concentration of solute ions or molecules in solution will have the lowest freezing point.

Solution	Particle solution
(a) 6.2 M NaCl	2 x 6.2 M = 12.4 M ions
(b) 2.1 M Al(NO$_3$)$_3$	4 x 2.1 M = 8.4 M ions
(c) 4.3 M K$_2$SO$_3$	3 x 4.3 M = 12.9 M ions

Solution (c) has the highest concentration of solute particles (ions), therefore it will have the lowest freezing point.

6.12 $310. \text{ g ethanol} \left(\dfrac{1 \text{ mol ethanol}}{46.07 \text{ g ethanol}} \right) = 6.73 \text{ mol ethanol}$

Aqueous ethanol does not dissociate into ions, therefore,

$\Delta T = \left(0.52 \dfrac{^\circ C}{mol} \right)(6.73 \text{ mol}) = 3.5 \ ^\circ C$, so the boiling point of the ethanol-water solution is 103.5 $^\circ$C.

6.13 Osmolarity = M x i
First, calculate the molarity (M):

$M_{Na_3PO_4} = \dfrac{3.3 \text{ g Na}_3\text{PO}_4}{100 \text{ mL sol}} \left(\dfrac{1 \text{ mol Na}_3\text{PO}_4}{163.9 \text{ g Na}_3\text{PO}_4} \right) \left(\dfrac{1000 \text{ mL sol}}{1 \text{ L sol}} \right) = 0.20 \ M \text{ Na}_3\text{PO}_4$

A Na$_3$PO$_4$ molecule gives 3 Na$^+$ ions and 1 PO$_4^{3-}$ ion for a total of 4 solute particles.
Osmolarity = (0.20 M)(4 ions) = 0.80 osmol

6.14 The osmolarity of red blood cells is 0.30 osmol:

Solution	Particle solution
(a) 0.1 M Na$_2$SO$_4$	3 x 0.1 M = 0.3 osmol
(b) 1.0 M Na$_2$SO$_4$	3 x 1.0 M = 3.0 osmol
(c) 0.2 M Na$_2$SO$_4$	3 x 0.2 M = 0.6 osmol

Solution (a) has the same osmolarity as red blood cells, therefore is isotonic compared to red blood cells.

6.15 (a), (b), and (d): True
(c) False: solutions cannot be separated by filtration.

6.17 Vinegar is an aqueous solution of acetic acid, where water is the solvent.

6.19 (a) Bronze contains both tin and copper as solids.
(b) Coffee contains solid solute (caffeine, flavorings) and liquid solvent (water).
(c) Car exhaust contains both CO_2 and steam (H_2O) as gases.
(d) Champaign contains CO_2 gas and liquid ethanol solutes in liquid water solvent.

6.21 Mixtures of gases, which mix in all proportions, are true solutions because their combinations are transparent, their molecules are distributed uniformly, and the component gases do not separate upon standing.

6.23 The prepared aspartic acid solution was unsaturated. Over two days time, some of the solvent (water) may have evaporated and the solution became supersaturated, precipitating the excess aspartic acid as a white solid.

6.25 (a) Ionic NaCl will dissolve in the polar water layer.
(b) Nonpolar camphor will dissolve in the nonpolar diethyl ether layer.
(c) Ionic KOH will dissolve in the polar layer.

6.27 Isopropyl alcohol would be a good first choice. The oil base in the paint is nonpolar. Both benzene and hexane are nonpolar solvents and may dissolve the paint, thus destroying the painting.

6.29 The solubility of aspartic acid at 25°C in 50.0 mL water is 0.250 g of solute. The cooled solution of 0.251 g aspartic acid in 50.0 mL water will be supersaturated by 0.001 g of aspartic acid.

6.31 According to Henry's Law, the solubility of a gas in a liquid is directly proportional to pressure. A closed bottle of a carbonated beverage is under pressure. After the bottle is opened, the pressure is released and the carbon dioxide becomes less soluble and escapes, leaving the contents "flat".

6.33 (a) Both quantities are equivalent to one significant figure.

$$\frac{1\ \cancel{min}}{2\ \cancel{yr}}\left(\frac{1\ \cancel{yr}}{365\ \cancel{days}}\right)\left(\frac{1\ \cancel{day}}{24\ \cancel{hr}}\right)\left(\frac{1\ \cancel{hr}}{60\ \cancel{min}}\right)\times 10^6\ ppm = 1\ ppm$$

$$\frac{1\ \cancel{cent}}{10,000\ \cancel{dol}}\left(\frac{1\ \cancel{dol}}{100\ \cancel{cents}}\right)\times 10^6\ ppm = 1\ ppm$$

(b) Both quantities are equivalent to one significant figure.

$$\frac{1\ \cancel{min}}{2000\ \cancel{yr}}\left(\frac{1\ \cancel{yr}}{365\ \cancel{days}}\right)\left(\frac{1\ \cancel{day}}{24\ \cancel{hr}}\right)\left(\frac{1\ \cancel{hr}}{60\ \cancel{min}}\right)\times 10^9\ ppb = 1\ ppb$$

$$\frac{1\ \cancel{cent}}{10,000,000\ \cancel{dol}}\left(\frac{1\ \cancel{dol}}{100\ \cancel{cents}}\right)\times 10^9\ ppb = 1\ ppb$$

6.35 (a) Vol. of ethanol $= 280\ \cancel{mL\ sol}\left(\frac{27\ mL\ ethanol}{100\ \cancel{mL\ sol}}\right)= 76\ mL\ ethanol$

76 mL ethanol dissolved in 204 mL water (to give 280 mL of solution)

(b) Vol. of ethyl acetate $= 435\ \cancel{mL\ sol}\left(\frac{1.8\ mL\ ethyl\ acetate}{100\ \cancel{mL\ sol}}\right)= 7.8\ mL\ ethyl\ acetate$

7.8 mL ethyl acetate dissolved in 427.2 mL water (to give 435 mL of solution)

(c) Vol. of benzene $= 1.65\ \cancel{L\ sol}\left(\frac{1000\ \cancel{mL\ sol}}{1\ \cancel{L\ sol}}\right)\left(\frac{8.00\ mL\ benzene}{100\ \cancel{mL\ sol}}\right)= 132\ mL\ benzene$

0.132 L benzene dissolved in 1.52 L chloroform (to give 1.65 L of solution)

6.37 (a) % (w/v) $= \frac{623\ \cancel{mg\ casein}}{15.0\ mL\ sol}\left(\frac{1\ g\ casein}{1000\ \cancel{mg\ casein}}\right)\times 100\% = 4.15\ \%\ w/v\ casein$

(b) % (w/v) $= \frac{74\ \cancel{mg\ vit.\ C}}{250\ mL\ sol}\left(\frac{1\ g\ vit.\ C}{1000\ \cancel{mg\ vit.\ C}}\right)\times 100\% = 0.030\ \%\ w/v\ vitamin\ C$

(c) % (w/v) $= \frac{3.25\ g\ sucrose}{186\ mL\ sol}\times 100\% = 1.75\ \%\ w/v\ sucrose$

6.39 (a) $175 \text{ mL sol} \left(\dfrac{1 \text{ L sol}}{1000 \text{ mL sol}} \right) \left(\dfrac{1.14 \text{ mol NH}_4\text{Br}}{1 \text{ L solution}} \right) \left(\dfrac{97.9 \text{ g NH}_4\text{Br}}{1 \text{ mol NH}_4\text{Br}} \right) = 19.5 \text{ g NH}_4\text{Br}$

Place 19.5 g of NH_4Br into a 175-mL volumetric flask, add some water, swirl until completely dissolved, and then fill the flask with water to the 175-mL mark.

(b) $1.35 \text{ L sol} \left(\dfrac{0.825 \text{ mol NaI}}{1 \text{ L solution}} \right) \left(\dfrac{149.9 \text{ g NaI}}{1 \text{ mol NaI}} \right) = 167 \text{ g NaI}$

Place 167 g of NaI into a 1.35-L volumetric flask, add some water, swirl until completely dissolved, and then fill the flask with water to the 1.35-L mark.

(c) $330 \text{ mL sol} \left(\dfrac{1 \text{ L sol}}{1000 \text{ mL sol}} \right) \left(\dfrac{0.16 \text{ mol ethanol}}{1 \text{ L solution}} \right) \left(\dfrac{46.1 \text{ g ethanol}}{1 \text{ mol ethanol}} \right) = 2.4 \text{ g ethanol}$

Place 2.4 g of ethanol into a 330-mL volumetric flask, add some water, swirl until completely mixed, and then fill the flask with water to the 330-mL mark.

6.41 $M_{NaCl} = \dfrac{5.0 \text{ mg NaCl}}{0.5 \text{ mL sol}} \left(\dfrac{1 \text{ g NaCl}}{1000 \text{ mg NaCl}} \right) \left(\dfrac{1 \text{ mol NaCl}}{58.4 \text{ g NaCl}} \right) \left(\dfrac{1000 \text{ mL sol}}{1 \text{ L sol}} \right) = 0.2 \ M \text{ NaCl}$

6.43 $M_{glucose} = \dfrac{22.0 \text{ g glucose}}{240. \text{ mL sol}} \left(\dfrac{1 \text{ mol glucose}}{180. \text{ g glucose}} \right) \left(\dfrac{1000 \text{ mL sol}}{1 \text{ L sol}} \right) = 0.509 \ M \text{ glucose}$

$M_{K+} = \dfrac{190. \text{ mg K}^+}{240. \text{ mL sol}} \left(\dfrac{1 \text{ g}}{1000 \text{ mg}} \right) \left(\dfrac{1 \text{ mol K}^+}{39.1 \text{ g K}^+} \right) \left(\dfrac{1000 \text{ mL sol}}{1 \text{ L sol}} \right) = 0.0202 \ M \text{ K}^+$

$M_{Na+} = \dfrac{4.00 \text{ mg K}^+}{240. \text{ mL sol}} \left(\dfrac{1 \text{ g}}{1000 \text{ mg}} \right) \left(\dfrac{1 \text{ mol Na}^+}{23.0 \text{ g Na}^+} \right) \left(\dfrac{1000 \text{ mL sol}}{1 \text{ L sol}} \right) = 7.25 \times 10^{-4} \ M \text{ Na}^+$

6.45 $M_{sucrose} = \dfrac{13 \text{ g sucrose}}{15 \text{ mL sol}} \left(\dfrac{1 \text{ mol sucrose}}{342.3 \text{ g sucrose}} \right) \left(\dfrac{1000 \text{ mL sol}}{1 \text{ L sol}} \right) = 2.5 \ M \text{ sucrose}$

6.47 Use the $\%_1 V_1 = \%_2 V_2$ equation:

$V_2 = \dfrac{(0.750\% \text{ w/v albumin})(5.00 \text{ mL sol})}{(0.125\% \text{ w/v albumin})} = 30.0 \text{ mL}$

The total volume of the dilution is 30.0 mL. Starting with a 5.00 mL solution, <u>25.0 mL of water must be added</u> to reach a final volume of 30.0 mL.

6.49 Use the $\%_1 V_1 = \%_2 V_2$ equation:

$$V_1 = \frac{\left(0.25\% \text{ w/v } H_2O_2\right)\left(250 \text{ mL sol}\right)}{\left(30.0\% \text{ w/v } H_2O_2\right)} = 2.1 \text{ mL } H_2O_2$$

Place 2.1 mL of 30.0% w/v H_2O_2 into a 250-mL volumetric flask, add some water, swirl until completely mixed, and then fill the flask with water to the 250-mL mark.

6.51 (a) $\dfrac{12.5 \text{ mg Captopril}}{325 \text{ mg pill}} \times 10^6 \text{ ppm} = 3.85 \times 10^4 \text{ ppm Captopril}$

(b) $\dfrac{22 \text{ mg Mg}^{2+}}{325 \text{ mg pill}} \times 10^6 \text{ ppm} = 6.8 \times 10^4 \text{ ppm Mg}^{2+}$

(c) $\dfrac{0.27 \text{ mg Ca}^{2+}}{325 \text{ mg pill}} \times 10^6 \text{ ppm} = 8.3 \times 10^2 \text{ ppm Ca}^{2+}$

6.53 Assume the density of the lake water to be 1.0 g/mL

$$1 \times 10^7 \text{ L water} \left(\frac{1000 \text{ mL water}}{1 \text{ L water}}\right)\left(\frac{1.0 \text{ g water}}{1.0 \text{ mL water}}\right) = 1 \times 10^{10} \text{ g water}$$

$$\frac{0.1 \text{ g dioxin}}{1 \times 10^{10} \text{ g water}} \times 10^9 \text{ ppb} = 0.01 \text{ ppb dioxin}$$

No, the dioxin level in the lake did not reach a dangerous level.

6.55 First, calculate the mass (in grams) of the nutrients in the cheese based on their percentages of daily allowances, then calculate concentration in ppm:

$$\text{Iron:} \quad (0.02)(15 \text{ mg Fe})\left(\frac{1 \text{ g Fe}}{1000 \text{ mg Fe}}\right) = 3 \times 10^{-4} \text{ g Fe}$$

$$\frac{3 \times 10^{-4} \text{ g Fe}}{28 \text{ g cheese}} \times 10^6 \text{ ppm} = 1 \times 10^1 \text{ ppm Fe}$$

$$\text{Calcium:} \quad (0.06)(1200 \text{ mg Ca})\left(\frac{1 \text{ g Ca}}{1000 \text{ mg Ca}}\right) = 7 \times 10^{-2} \text{ g Ca}$$

$$\frac{7 \times 10^{-2} \text{ g Ca}}{28 \text{ g cheese}} \times 10^6 \text{ ppm} = 3 \times 10^3 \text{ ppm Ca}$$

$$\text{Vitamin A:} \quad (0.06)(0.800 \text{ mg Vit. A})\left(\frac{1 \text{ g}}{1000 \text{ mg Vit. A}}\right) = 5 \times 10^{-5} \text{ g Vitamin A}$$

$$\frac{5 \times 10^{-5} \text{ g Vit. A}}{28 \text{ g cheese}} \times 10^6 \text{ ppm} = 2 \text{ ppm Vitamin A}$$

6.57 (a) Ionic compounds completely dissociate in water: strong electrolyte.
(b) Covalently bonded compounds do not dissociate in water: non-electrolyte.
(c) Strong bases completely dissociate in water: strong electrolyte.
(d) Weak acids only partially dissociate in water: weak electrolyte.
(e) Covalently bonded compounds do not dissociate in water: non-electrolyte.

6.59 Ethanol and water are infinitely miscible. Ethanol dissolves in water by forming hydrogen bonds with water.

6.61 (a) and (b) are both true

6.63 (a) Homogeneous   (b) Heterogeneous   (c) Colloidal
(d) Heterogeneous   (e) Colloidal   (f) Colloidal

6.65 As the temperature of the solution decreased, the protein molecules must have aggregated and formed a colloidal mixture. The turbid appearance is the result of the Tyndall effect.

6.67 $\Delta T = (1.86°C/mol)$(mole of particles in solution per 1000. g of water):

(a) $\Delta T = 1.00$ ~~mol NaCl~~ $\left(\dfrac{2\text{ mol particle}}{1\text{ mol NaCl}}\right)\left(\dfrac{1.86°C}{\text{mol particle}}\right) = 3.72°C$ f.p. = -3.72°C

(b) $\Delta T = 1.00$ ~~mol MgCl₂~~ $\left(\dfrac{3\text{ mol particle}}{1\text{ mol MgCl}_2}\right)\left(\dfrac{1.86°C}{\text{mol particle}}\right) = 5.58°C$ f.p. = -5.58°C

(c) $\Delta T = 1.00$ ~~mol (NH₄)₂CO₃~~ $\left(\dfrac{3\text{ mol particle}}{1\text{ mol }(NH_4)_2CO_3}\right)\left(\dfrac{1.86°C}{\text{mol particle}}\right) = 5.58°C$

f.p. = -5.58°C

(d) $\Delta T = 1.00$ ~~mol Al(HCO₃)₃~~ $\left(\dfrac{4\text{ mol particle}}{1\text{ mol Al(HCO}_3)_3}\right)\left(\dfrac{1.86°C}{\text{mol particle}}\right) = 7.44°C$

f.p. = -7.44°C

6.69 Methanol is a non-electrolyte and does not dissociate. Use the following equation for freezing point depression while also converting to grams.

Moles of CH_3OH per 1000. g $H_2O = \Delta T\left(\dfrac{1\text{ mol particles}}{1.86°C}\right)$

$20.°C\left(\dfrac{1\text{ mol particles}}{1.86°C}\right)\left(\dfrac{1\text{ mole CH}_3\text{OH}}{1\text{ mol particles}}\right)\left(\dfrac{32.0\text{ g CH}_3\text{OH}}{1\text{ mol CH}_3\text{OH}}\right) = 3.4 \times 10^2$ g CH_3OH

6.71 Acetic acid is a weak acid, therefore does not completely dissociate into ions. KF is a strong electrolyte, completely dissociating into two ions and doubling the effect on freezing point depression compared to acetic acid.

6.73 In each case, the side with the greater osmolarity rises.
(a) B   (b) B   (c) A   (d) B   (e) A   (f) neither

6.75 (a) $\text{osmol}_{\text{Na}_2\text{CO}_3} = \dfrac{0.39 \text{ mol Na}_2\text{CO}_3}{1 \text{ L sol}}\left(\dfrac{3 \text{ mol particles}}{1 \text{ mol Na}_2\text{CO}_3}\right) = 1.2 \text{ osmol}$

(b) $\text{osmol}_{\text{Al(NO}_3)_3} = \dfrac{0.62 \text{ mol Al(NO}_3)_3}{1 \text{ L sol}}\left(\dfrac{4 \text{ mol particles}}{1 \text{ mol Al(NO}_3)_3}\right) = 2.5 \text{ osmol}$

(c) $\text{osmol}_{\text{LiBr}} = \dfrac{4.2 \text{ mol LiBr}}{1 \text{ L sol}}\left(\dfrac{2 \text{ mol particles}}{1 \text{ mol LiBr}}\right) = 8.4 \text{ osmol}$

(d) $\text{osmol}_{K_3PO_4} = \dfrac{0.009 \text{ mol K}_3\text{PO}_4}{1 \text{ L sol}}\left(\dfrac{4 \text{ mol particles}}{1 \text{ mol K}_3\text{PO}_4}\right) = 0.04 \text{ osmol}$

6.77 Cells in hypertonic solutions undergo crenation (shrink).

$\text{osmol}_{\text{NaCl}} = \dfrac{0.9 \text{ g NaCl}}{100 \text{ mL sol}}\left(\dfrac{1000 \text{ mL sol}}{1 \text{ L sol}}\right)\left(\dfrac{1 \text{ mol NaCl}}{58.4 \text{ g NaCl}}\right)\left(\dfrac{2 \text{ mol particles}}{1 \text{ mol NaCl}}\right) = 0.3 \text{ osmol}$

(a) $\dfrac{0.3 \text{ g NaCl}}{100 \text{ mL sol}}\left(\dfrac{1000 \text{ mL sol}}{1 \text{ L sol}}\right)\left(\dfrac{1 \text{ mol NaCl}}{58.4 \text{ g NaCl}}\right)\left(\dfrac{2 \text{ mol particles}}{1 \text{ mol NaCl}}\right) = 0.1 \text{ osmol NaCl}$

(b) $\text{osmol}_{\text{glucose}} = 0.9 \ M \times 1 \text{ particle} = 0.9 \text{ osmol}$

(c) $\dfrac{0.9 \text{ g glu}}{100 \text{ mL sol}}\left(\dfrac{1000 \text{ mL sol}}{1 \text{ L sol}}\right)\left(\dfrac{1 \text{ mol glu}}{180 \text{ g glu}}\right)\left(\dfrac{1 \text{ mol particles}}{1 \text{ mol glu}}\right) = 0.05 \text{ osmol glucose}$

Solution (b) has a concentration greater than the isotonic solution so it will crenate red blood cells.

6.79 Carbon dioxide (CO_2) dissolves in normal rainwater to form a dilute solution of carbonic acid (H_2CO_3), which is a weak acid.

6.81 Deep-sea divers must breathe compressed air under high pressure. The nitrogen in air can dissolve in the blood from pressures experienced at depths greater than 40 meters. Nitrogen dissolved in the blood can lead to nitrogen narcosis, a narcotic effect also referred to as "rapture of the deep", which is similar to alcohol-induced intoxication. Helium does not induce this effect so an oxygen-helium mixture is used instead.

6.83 $250 \text{ mL sol}\left(\dfrac{1 \text{ L sol}}{1000 \text{ mL sol}}\right)\left(\dfrac{4.6 \text{ mEq Ca}^{2+}}{1 \text{ L sol}}\right)\left(\dfrac{1 \text{ mmol Ca}^{2+}}{2 \text{ mEq}}\right) = 5.75 \times 10^{-1} \text{ mmol Ca}^{2+}$

$0.575 \text{ mmol Ca}^{2+}\left(\dfrac{1 \text{ mol Ca}^{2+}}{1000 \text{ mmol Ca}^{2+}}\right)\left(\dfrac{40.08 \text{ g Ca}^{2+}}{1 \text{ mol Ca}^{2+}}\right)\left(\dfrac{1000 \text{ mg Ca}^{2+}}{1 \text{ g Ca}^{2+}}\right) = 23 \text{ mg Ca}^{2+}$

6.85 $CaCO_3 + H_2SO_4 \rightarrow CaSO_4 + CO_2 + H_2O$

$CaSO_4 + 2H_2O \rightarrow CaSO_4 \cdot 2H_2O$

6.87 The minimum pressure required for the reverse osmosis in the desalinization of seawater exceeds 100 atm (the osmotic pressure of sea water).

6.89 $\dfrac{0.2 \text{ g NaHCO}_3}{100 \text{ mL sol}}\left(\dfrac{1000 \text{ mL sol}}{1 \text{ L sol}}\right)\left(\dfrac{1 \text{ mol NaHCO}_3}{84.0 \text{ g NaHCO}_3}\right)\left(\dfrac{2 \text{ mol particles}}{1 \text{ mol NaHCO}_3}\right) = 0.05 \text{ osmol}$

$\text{osmol}_{NaHCO_3} = 0.05 \text{ osmol}$

$\dfrac{0.2 \text{ g KHCO}_3}{100 \text{ mL sol}}\left(\dfrac{1000 \text{ mL sol}}{1 \text{ L sol}}\right)\left(\dfrac{1 \text{ mol KHCO}_3}{100.1 \text{ g KHCO}_3}\right)\left(\dfrac{2 \text{ mol particles}}{1 \text{ mol KHCO}_3}\right) = 0.04 \text{ osmol}$

$\text{osmol}_{KHCO_3} = 0.04 \text{ osmol}$

Yes, the change made a change in the tonicity. The error in replacing $NaHCO_3$ with $KHCO_3$ resulted in a hypotonic solution and an electrolyte imbalance by reducing the number of ions (osmolarity) in solution.

6.91 When the cucumber is placed in a saline solution, the osmolarity of the saline is greater than the water in the cucumber, so water moves from the cucumber to the saline solution. When a prune (partially dehydrated plum) is placed in the same solution, it expands because the osmolarity inside the prune is greater than the saline solution, so the water moves from saline solution to inside the prune.

6.93 The solubility of a gas is directly proportional to the pressure (Henry's Law) and inversely proportional to the temperature. The dissolved carbon dioxide formed a saturated solution in water when bottled at 2 atm of pressure. When the bottles are opened at atmospheric pressure, the gas becomes less soluble in the water. The carbon dioxide becomes supersaturated in water at room temperature and 1 atm, thus escapes through bubbles and frothing. In the other bottle, the solution of carbon dioxide in water is unsaturated at lower temperatures and does not lose carbon dioxide.

6.95 Methanol is more efficient at lowering the freezing point of water because given mass of methanol (32 g/mol) contains a greater number of moles than the same mass of ethylene glycol (62 g/mol).

6.97 $CO_2(g) + H_2O(l) \rightarrow H_2CO_3(aq)$ carbonic acid

$SO_2(g) + H_2O(g) \rightarrow H_2SO_3(aq)$ sulfurous acid

6.99 Use the $\%_1 V_1 = \%_2 V_2$ equation:

$$V_1 = \frac{\left(4.5\% \text{ w/v } HNO_3\right)\left(300 \text{ mL sol}\right)}{\left(35\% \text{ w/v } HNO_3\right)} = 39 \text{ mL } HNO_3$$

Place 39 mL of 35% w/v HNO_3 into a 300-mL volumetric flask, add some water, swirl until completely mixed, and then fill the flask with water to the 300-mL mark.

6.101 $1016 \text{ kg } H_2O \left(\dfrac{1000 \text{ g } H_2O}{1 \text{ kg } H_2O}\right)\left(\dfrac{6 \text{ g pollutant}}{10^9 \text{ g } H_2O}\right) = 6 \times 10^{-3} \text{ g pollutant}$

6.103 Assume that the density of the pool water is 1.00 g/mL

$$[Cl_2] = \frac{0.00500 \text{ mol } Cl_2}{1 \text{ L sol}}\left(\frac{70.9 \text{ g } Cl_2}{1 \text{ mol } Cl_2}\right)\left(\frac{1 \text{ L sol}}{1000 \text{ mL sol}}\right)\left(\frac{1 \text{ mL sol}}{1 \text{ g sol}}\right) \times 10^6 \text{ ppm} = 355 \text{ ppm}$$

$$20,000. \text{ L } H_2O \left(\frac{0.00500 \text{ mol } Cl_2}{1 \text{ L } H_2O}\right)\left(\frac{70.9 \text{ g } Cl_2}{1 \text{ mol } Cl_2}\right)\left(\frac{1 \text{ kg } Cl_2}{1000 \text{ g } Cl_2}\right) = 7.09 \text{ kg } Cl_2 \text{ added}$$

6.105 Assuming that the concentration is 10.0% w/w H_2SO_4:

$$8.37 \text{ g } H_2SO_4 \left(\frac{100 \text{ g sol}}{10.0 \text{ g } H_2SO_4}\right)\left(\frac{1 \text{ mL sol}}{1.07 \text{ g sol}}\right) = 78.2 \text{ mL of } 10\% \text{ } H_2SO_4 \text{ sol}$$

6.107 (a) $25.0 \text{ g } MgCl_2 \left(\dfrac{1 \text{ mol } MgCl_2}{95.21 \text{ g } MgCl_2}\right)\left(\dfrac{3 \text{ mol ions}}{1 \text{ mol } MgCl_2}\right) = 0.788 \text{ mol particles}$

$$\Delta T_f = \frac{-1.86°C}{mol} \times \text{mol of particles} = \frac{-1.86°C}{mol} \times 0.788 \text{ mol} = -1.47°C$$

new freezing point = -1.47°C

(b) $\Delta T_b = \dfrac{0.512°C}{mol} \times \text{mol of particles} = \dfrac{0.512°C}{mol} \times 0.788 \text{ mol} = 0.403°C$

new boiling point = 100.403°C

6.109 First, balance the chemical equation:

$$Ca + 2HBr \rightarrow CaBr_2 + H_2$$

(a) $1.46 \text{ g Ca} \left(\dfrac{1 \text{ mol Ca}}{40.08 \text{ g Ca}} \right) \left(\dfrac{1 \text{ mol } H_2}{1 \text{ mol Ca}} \right) = 3.64 \times 10^{-2} \text{ mol } H_2$

$115 \text{ mL HBr sol} \left(\dfrac{1 \text{ L sol}}{1000 \text{ mL sol}} \right) \left(\dfrac{0.325 \text{ mol HBr}}{1 \text{ L sol}} \right) \left(\dfrac{1 \text{ mol } H_2}{2 \text{ mol HBr}} \right) = 1.87 \times 10^{-2} \text{ mol } H_2$

HBr is the limiting reagent

(b)   First, convert the units to those required by the ideal gas law equation:

Dalton's law of partial pressures: $P_T = P_{H2} + P_{H2O}$

$P_{H2} = P_T - P_{H2O} = 754 \text{ torr} - 21 \text{ torr} = 733 \text{ torr}$

$P_{H2} = 733 \text{ torr} \left(\dfrac{1.00 \text{ atm}}{760. \text{ torr}} \right) = 0.964 \text{ atm}$ $T(K) = 22 + 273 = 295 \text{ K}$

Now, plug the appropriate values into the ideal gas law equation:

$$V = \dfrac{nRT}{P} = \dfrac{(0.0187 \text{ mol})(0.0821 \text{ L} \cdot \text{atm} \cdot \text{mol}^{-1} \cdot K^{-1})(295 K)}{0.964 \text{ atm}} = 0.470 \text{ L } H_2$$

(c) $1.46 \text{ g Ca} \left(\dfrac{1 \text{ mol Ca}}{40.08 \text{ g Ca}} \right) = 3.64 \times 10^{-2} \text{ mol Ca}$

$115 \text{ mL HBr sol} \left(\dfrac{1 \text{ L sol}}{1000 \text{ mL sol}} \right) \left(\dfrac{0.325 \text{ mol HBr}}{1 \text{ L sol}} \right) \left(\dfrac{1 \text{ mol Ca}}{2 \text{ mol HBr}} \right) = 1.87 \times 10^{-2} \text{ mol Ca}$

$3.64 \times 10^{-2} \text{ mol Ca (start)} - 1.87 \times 10^{-2} \text{ (reacted)} = 1.77 \times 10^{-2} \text{ mol Ca left over}$

$1.77 \times 10^{-2} \text{ mol Ca} \left(\dfrac{40.08 \text{ g Ca}}{1 \text{ mol Ca}} \right) = 0.709 \text{ g Ca left unreacted}$

Chapter 7: Reaction Rates and Chemical Equilibrium

7.1 The rate of the reaction is equal to the change in the amount of O_2 per unit time.

$$\text{Rate of } O_2 \text{ formation} = \frac{\left(0.35 \text{ L } O_2 - 0.020 \text{ L } O_2\right)}{15 \text{ min}} = 0.022 \text{ L } O_2/\text{min}$$

7.2 Rate = $k[H_2O_2]$ = $(0.01/\text{min})(0.36 \text{ mol } H_2O_2/\text{L})$ = 4×10^{-3} mol $H_2O_2/\text{L} \cdot \text{min}$ for the disappearance of H_2O_2.

7.3 $K = \dfrac{\left[H_2SO_4\right]}{\left[SO_3\right]\left[H_2O\right]}$

7.4 $K = \dfrac{\left[N_2\right]\left[H_2\right]^3}{\left[NH_3\right]^2}$

7.5 $K = \dfrac{\left[PCl_5\right]}{\left[PCl_3\right]\left[Cl_2\right]} = \dfrac{\left[1.66 \, M\right]}{\left[1.66 \, M\right]\left[1.66 \, M\right]} = 0.602 \, M^{-1}$

7.6 Le Chatelier's principle predicts that the equilibrium would shift to the left (reactants) by adding a product (Br_2).

7.7 Because oxygen's solubility in water has been exceeded, oxygen bubbles out of solution, driving the equilibrium towards the right (products).

7.8 If the equilibrium shifts right with the addition of heat, heat must have been a reactant and the reaction endothermic.

7.9 Reaction equilibria where there is an increase in pressure favor the side with fewer moles of gas, therefore the equilibrium in question shifts to the right.

7.10 rate = $\dfrac{\Delta[HCl]}{\Delta time}$ = $\dfrac{0.27M}{54.0 \text{ min}}$ = 5.0×10^{-3} M/min

7.11 In the solution of this problem, it is assumed that 1 hour and 20 minutes is equal to 80. minutes, with two significant figures.

$$\text{Rate of } CH_3I \text{ formation} = \frac{\left(0.840 \, M \, CH_3I - 0.260 \, M \, CH_3I\right)}{80 \text{ min}} = 7.3 \times 10^{-3} \, M \, CH_3I/\text{min}$$

7.13 Reactions involving ions in aqueous solution are faster because they require no covalent bond breaking and have low activation energies. In addition, the attractive forces between oppositely charged ions provide energy to drive the reaction. Reactions between covalent molecules involve breaking covalent bonds, requiring higher activation energies that result in slower reaction rates.

7.15 The following energy diagram can be drawn for an exothermic reaction:

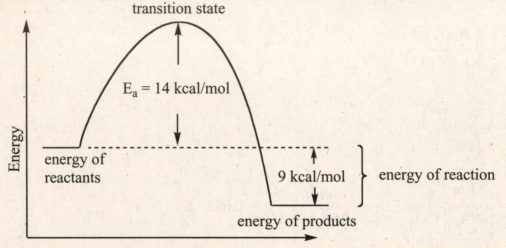

7.17 The general rule for temperature effect on reaction rates states that for every temperature increase of 10 °C, the reaction rate doubles.

Temperature: 10°C → 20°C → 30°C → 40°C → 50°C
Rate: 16 hr → 8 hr → 4 hr → 2 hr → 1 hr

A reaction temperature of 50 °C corresponds to a reaction completion time of 1 hr.

7.19 (1) Increase the temperature
(2) Increase the concentration of reactants
(3) Add a catalyst

7.21 A catalyst increases the rate by providing an alternative reaction pathway with lower activation energy.

7.23 Examples of irreversible reactions include: digesting a piece of candy, the rusting of iron, exploding TNT, and the reaction of sodium or potassium with water.

7.25 (a) $K = \dfrac{[H_2O]^2[O_2]}{[H_2O_2]^2}$

(b) $K = \dfrac{[N_2O_4]^2[O_2]}{[N_2O_5]^2}$

(c) $K = \dfrac{[C_6H_{12}O_6][O_2]^6}{[H_2O]^6[CO_2]^6}$

7.27 $K = \dfrac{[CO_2][H_2]}{[H_2O][CO]} = \dfrac{[0.133\ M][3.37\ M]}{[0.720\ M][0.933\ M]} = 0.667$

7.29 $K = \dfrac{[NO]^2[Cl_2]}{[NOCl_2]^2} = \dfrac{[1.4\ M]^2[0.34\ M]}{[2.6\ M]^2} = 0.099\ M$

7.31 When K > 1, equilibrium favors products; when K < 1, equilibrium favors reactants. Products are favored in (b) and (c). Reactants are favored in (a), (d), and (e)

7.33 No, the rate of reaction is independent of the energy difference between products and reactants. The rate of reaction is inversely proportional to the activation energy.

7.35 The reaction reaches equilibrium quickly, but the equilibrium favors the reactants (K < 1) and would not be a very good industrial process.

7.37 (a) Right   (b) Right   (c) Left   (d) Left   (e) No shift

7.39 (a) Adding Br_2 (a reactant) will shift the equilibrium to the right.
(b) The equilibrium constant will remain the same.

7.41 (a) No change   (b) No change   (c) Smaller
Equilibrium constants are independent of reactant and product concentrations. The K of an endothermic reaction will decrease with decreasing temperature

7.43 As temperatures increase, the rates of most chemical processes increase. A high body temperature is dangerous because metabolic processes (including digestion, respiration, and the biosynthesis of essential compounds) take place at a rate faster than what is safe for the body. As temperatures decrease, so do the rates of most chemical reactions. As body temperatures decrease below normal, the vital chemical reactions will slow down to rates slower than what is safe for the body.

7.45 The capsule with the tiny beads will act faster than the solid pill form. The small bead size increases the drug's surface area allowing the drug to react faster and deliver its therapeutic effects more quickly.

7.47 Assuming that there is excess AgCl in the previous recipe, the new recipe does not need to change. The desert conditions add nothing that would affect an equilibrium initiated by light.

7.49 If the rate of reaction at 15 °C is 2.8 moles HCl per liter per second, at –5 °C (a decrease of 20 °C), the rate will decrease twice by one-half (for a total of 1/4) to 0.70 moles HCl per liter per second. If the temperature is increased from 15 °C to 45 °C (an increase of 30 °C), the rate will double three times to 22 moles HCl per liter per second.

7.51 The difference between profiles **1** and **2** is that energy profile **2** occurred via an alternative pathway involving a lower activation energy accomplished by a catalyst.

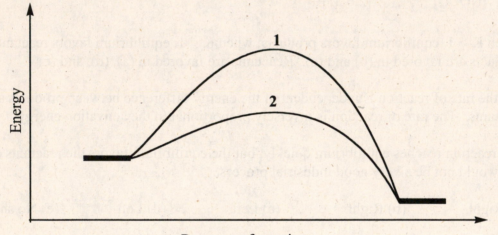

7.53 $K = \dfrac{[NO]^2[Br_2]}{[NOBr]^2} = 25\,M$

$[NOBr] = \sqrt{[NOBr]^2} = \sqrt{\dfrac{[0.80\,M]^2[0.80\,M]}{25\,M}} = 0.14\,M$

7.55 For reactions in the gas phase, an increase in pressure or decrease in reaction container volume usually increases the reaction rate by increasing the number of collisions between molecules.

<u>7.57</u> $4NH_3(g) + 7O_2(g) \rightleftharpoons 4NO_2(g) + 6H_2O(g)$

<u>7.59</u> Reaction A involving spherical molecules goes faster because for spherical molecules, orientation does not matter for molecular collisions to yield product. Rod-like molecules require special orientations for effective collisions to occur.

<u>7.61</u> Rate $= \dfrac{[0.30 \text{ mol/L } I_2 - 0 \text{ mol/L } I_2]}{10. \text{ s}} = 3.0 \times 10^{-2} \text{ mol } I_2 / L \cdot s$

<u>7.63</u> (a)

	CH_3COOH	$\rightleftharpoons$	H^+	$+$	CH_3COO^-
initial concentration	0.01 M		0 M		0 M
change	- x		x		x
final concentration	0.01M – x		0 M + x		0 M + x

If the final concentration is 0.098M, it means x = 0.002 M (ignoring the significant figure implications). Therefore, [H+] = [CH₃COO⁻] = 0.002 M at equilibrium.

(b) $K_{eq} = \dfrac{[H^+][CH_3COO^-]}{[CH_3COOH]} = \dfrac{[0.002 \text{ M}][0.002 \text{ M}]}{[0.098 \text{ M}]} = 4 \times 10^{-5} \text{ M}$

<u>7.65</u> Monitoring the disappearance of a reactant is a possible way to determine the rate of a reaction. It will do just as well as monitoring the formation of product because the stoichiometry of the reaction relates the concentrations of products and reactants to one another.

<u>7.67</u> Some reactions are so fast that they are faster than the usual methods used to follow chemical reactions. Often, specialized equipment is necessary with sophisticated electronics fast enough to monitor very rapid reactions.

<u>7.69</u> The knowledge that a reaction is exothermic tells us nothing about the rate of reaction. The activation energy of this conversion is high, and therefore, the rate of conversion of diamond to graphite is so slow that it does not take place in any measurable amount of time.

<u>7.71</u> When sodium chloride is added, the presence of additional chloride ions increases one of the products of the equilibrium. The equilibrium then shifts to the left, increasing the amount of solid silver chloride.

7.73 $K_{eq} = \dfrac{[HF]^2}{[H_2][F_2]}$

$[HF] = \sqrt{[H_2][F_2]K_{eq}} = \sqrt{2.4 \times 10^{-3}[0.0021\ M][0.0021\ M]} = 1.0 \times 10^{-4}\ M$

7.75 **Increase pressure:** increases the concentration of NO_3^- because the equilibrium favors the side with fewer molecules of gas (reactant side).
Adding zinc: no change in the concentration of NO_2
Decrease the H^+: decreases the concentration of Zn^{+2}
Add Pt catalyst: no effect because catalysts do not affect the equilibrium concentrations, but only increase the rates that the equilibrium is established.
Add some Ar: assuming there is no pressure change; there is no change to the equilibrium concentrations because inert gases have no effect on equilibrium.
Decrease of Zn^{2+}: will have no effect on Keq. The concentrations will change to maintain the equilibrium constant.
Increase the temperature: the reaction is exothermic, therefore, shifts to the reactant side of the equilibrium, increasing the amount of zinc.

7.77 The activation energy of the reverse reaction is 10 kJ.

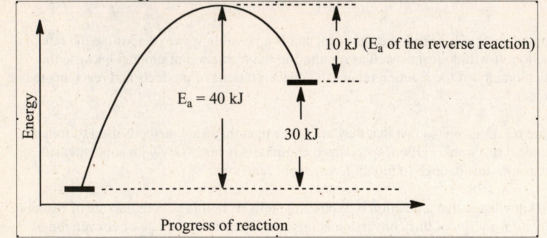

7.79 $2H_2O(g) \rightleftharpoons 2H_2(g) + O_2(g)$ $K_{eq} = \dfrac{[H_2]^2[O_2]}{[H_2O]^2} = 8.7 \times 10^3$

$\sqrt{[H_2O]^2} = \sqrt{\dfrac{[H_2]^2[O_2]}{8.7 \times 10^3}}$

$[H_2O] = \sqrt{\dfrac{[H_2]^2[O_2]}{8.7 \times 10^3}} = \sqrt{\dfrac{[1.9 \times 10^{-2}]^2[8.0 \times 10^{-2}]}{8.7 \times 10^3}} = 5.8 \times 10^{-5}\ M$

Chapter 8: Acids and Bases

8.1   Acid reaction for HPO_4^{2-}:

$$HPO_4^{2-} + H_2O \rightleftharpoons H_3O^+ + PO_4^{3-}$$

Base reaction for HPO_4^{2-}:

$$HPO_4^{2-} + H_2O \rightleftharpoons HO^- + H_2PO_4^{1-}$$

8.2   The following are acid-base equilibria:
(a) Equilibrium is left favored:

H_3O^+	+	I^-	$\longleftarrow\!\!\rightharpoonup$	H_2O	+	HI
weaker acid		weaker base		stronger base		stronger acid

(b) Equilibrium is left favored:

CH_3COO^-	+	H_2S	$\longleftarrow\!\rightharpoonup$	CH_3COOH	+	HS^-
weaker base		weaker acid		stronger acid		stronger base

8.3 $pK_a = -\log K_a$
$pK_a = -\log 4.9 \times 10^{-10} = 9.31$

8.4   The lower pK_a is the stronger acid:
(a) Ascorbic acid (vitamin C), pK_a 4.1
(b) Aspirin, pK_a 3.49

8.5 $[H_3O^+][OH^-] = 1.0 \times 10^{-14}$

$$[H_3O^+] = \frac{1.0 \times 10^{-14}}{1.0 \times 10^{-12}} = 1.0 \times 10^{-2}\ M$$

8.6 $pH = -\log [H_3O^+]$ $\qquad [H_3O^+] = 10^{-pH}$
(a) $[H_3O^+] = 3.5 \times 10^{-3}$ M; $pH = -\log[3.5 \times 10^{-3}\ M] = 2.46$
(b) pH of tomato juice is 4.10; $[H_3O^+] = 10^{-4.10} = 7.9 \times 10^{-5}\ M$. A solution with a pH less than 7 is acidic; therefore, tomato juice is acidic.

8.7 $pH + pOH = 14.00$
$[OH^-] = 1.0 \times 10^{-4}$ $\qquad pOH = 4.00$
$[H_3O^+] = 1.0 \times 10^{-10}$ $\qquad pH = 10.00$

8.8 The 25.0 mL acetic acid sample was titrated with an average volume of 19.83 mL 0.121 M NaOH solution.

$$19.83 \ \cancel{\text{mL NaOH}} \left(\frac{1 \ \cancel{\text{L NaOH}}}{1000 \ \cancel{\text{mL NaOH}}} \right) \left(\frac{0.121 \ \text{mol NaOH}}{1 \ \cancel{\text{L NaOH}}} \right) = 2.40 \times 10^{-3} \ \text{mol NaOH}$$

$$\text{Moles of acetic acid} = 2.40 \times 10^{-3} \ \cancel{\text{mol NaOH}} \left(\frac{1 \ \text{mol HOAc}}{1 \ \cancel{\text{mol NaOH}}} \right) = 2.40 \times 10^{-3} \ \text{mol}$$

$$[\text{Acetic acid}] = \frac{2.40 \times 10^{-3} \ \text{mol acetic acid}}{25.0 \ \cancel{\text{mL solution}}} \left(\frac{1000 \ \cancel{\text{mL sol}}}{1 \ \text{L sol}} \right) = 0.0960 \ M \ \text{acetic acid}$$

8.9 The pH of a buffer solution containing equimolar quantities of acid and conjugate base is equal to the pK_a of the weak acid.
(a) NH_4Cl and NH_3: pK_a for $NH_4^+ = 9.25$
(b) CH_3COOH and CH_3COONa: pK_a for $CH_3COOH = 4.74$

8.10 Use the Henderson-Hasselbalch Equation: $pH = pK_a + \log[A^-]/[HA]$ where: $pK_a = -\log K_a$ of the weak acid, HA and A^- are the weak acid and conjugate base respectively.

$$pH = -\log(4.0 \times 10^{-10}) + \log \left(\frac{0.50 \ M}{0.25 \ M} \right) = 9.70$$

8.11 Use the Henderson-Hasselbalch Equation: $pH = pK_a + \log[A^-]/[HA]$ where:

$$pH = 8.3 + \log \left(\frac{0.05 \ \text{mol}}{0.2 \ \text{mol}} \right) = 7.7$$

The mole amount of TRIS acid and base was used instead of concentration because they are both dissolved in 500 mL of water, therefore, the amount of water cancels out. The TRIS buffer will have a pH = 7.7.

8.13 Listed below are the following acid ionization equilibrium equations:

(a) $HNO_3(aq) + H_2O(l) \rightleftharpoons H_3O^+(aq) + NO_3^-(aq)$

(b) $HBr(aq) + H_2O(l) \rightleftharpoons H_3O^+(aq) + Br^-(aq)$

(c) $H_2SO_3(aq) + H_2O(l) \rightleftharpoons H_3O^+(aq) + HSO_3^-(aq)$

(d) $H_2SO_4(aq) + H_2O(l) \rightleftharpoons H_3O^+(aq) + HSO_4^-(aq)$

(e) $HCO_3^-(aq) + H_2O(l) \rightleftharpoons H_3O^+(aq) + CO_3^{2-}(aq)$

(f) $NH_4^+(aq) + H_2O(l) \rightleftharpoons H_3O^+(aq) + NH_3(aq)$

8.15 Acids (a), (c), (e), and (f) have pK_a's > 0, therefore do not ionize completely and are considered to be weak acids. Acids (b) and (d) have pK_a's < 0; therefore are ionized completely and are considered to be strong acids.

8.17 (c), (e), and (f): True
(a) False: acids with pK_a's > 0 are weak acids.
(b) False: acetic acid does not completely ionize, therefore, a 0.1 M solution has a pH >1.
(d) False: [H_3O^+] depends on the strength of the acid and its initial concentration.
(g) False: ammonia is a weak base.
(h) False: carbonic acid is a weak acid.

8.19 (a) Brønsted-Lowry acids are considered proton donors.
(b) Brønsted-Lowry bases are considered proton acceptors.

8.21 A conjugate base is the species that results after the acid lost its proton.
(a) HPO_4^{2-} (b) HS^- (c) CO_3^{2-}
(d) $CH_3CH_2O^-$ (e) OH^-

8.23 A conjugate acid results from a base acquiring a proton from an acid:
(a) H_3O^+ (b) $H_2PO_4^-$ (c) $CH_3NH_3^+$
(d) HPO_4^{2-} (e) NH_4^+

8.25 The equilibrium favors the side with the weaker acid-weaker base. Equilibria (b) and (c) favor the left, equilibrium (a) favors the right.

(a) C_6H_5OH + $C_2H_5O^-$ ⇌ $C_6H_5O^-$ + C_2H_5OH
stronger acid   stronger base   weaker base   weaker acid

(b) HCO_3^- + H_2O ⇌ H_2CO_3 + OH^-
weaker base   weaker acid   stronger acid   stronger base

(c) CH_3COOH + $H_2PO_4^-$ ⇌ CH_3COO^- + H_3PO_4
weaker acid   weaker base   stronger base   stronger acid

8.27 (a) Strong acids have smaller pK_a's; therefore weak acids have large pK_a's.
(b) Strong acids have large K_a's.

8.29 At equal concentrations, pH decreases (becomes more acidic) as K_a increases.
(a) 0.10 M HCl (b) 0.10 M H_3PO_4 (c) 0.010 M H_2CO_3
(d) 0.10 M NaH$_2$PO$_4$ (e) 0.10 M Aspirin

8.31 Only (b) involves a redox reaction. The other reactions are acid-base reactions.

(a) $Na_2CO_3 + 2HCl \rightarrow CO_2 + 2NaCl + H_2O$

(b) $Mg + 2HCl \rightarrow MgCl_2 + H_2$

(c) $NaOH + HCl \rightarrow NaCl + H_2O$

(d) $Fe_2O_3 \quad 6HCl \rightarrow 2FeCl_3 + 3H_2O$

(e) $NH_3 + HCl \rightarrow NH_4Cl$

(f) $CH_3NH_2 + HCl \rightarrow CH_3NH_3Cl$

(g) $NaHCO_3 + HCl \rightarrow H_2CO_3 + NaCl \rightarrow CO_2 + H_2O + NaCl$

8.33 Using the equation: $[H_3O^+][OH^-] = 1.0 \times 10^{-14} \, M^2$

(a) $[OH^-] = 10^{-3} \, M$ (b) $[OH^-] = 10^{-10} \, M$
(c) $[OH^-] = 10^{-7} \, M$ (d) $[OH^-] = 10^{-15} \, M$

8.35 Use the equation pH = -log$[H_3O^+]$:

(a) pH = 8 (basic)   (b) pH = 10 (basic)   (c) pH = 2 (acidic)
(d) pH = 0 (acidic)   (e) pH = 7 (neutral)

8.37 Use the equations pH = -log$[H_3O^+]$ and pH + pOH = 14.00.

(a) pH = 8.52 (basic)   (b) pH = 1.22 (acidic)
(c) pH = 11.10 (basic)   (d) pH = 6.30 (acidic)

8.39 Use the equations $[OH^-] = 10^{-pOH}$ and pH + pOH = 14.00.

(a) pOH = 1.00, $[OH^-]$ = 0.10 M   (b) pOH = 2.4, $[OH^-]$ = 4 x 10^{-3} M
(c) pOH = 2.0, $[OH^-]$ = 1 x 10^{-2} M   (d) pOH = 5.6, $[OH^-]$ = 3 x 10^{-6} M

8.41 $M = \dfrac{mol}{L} = \left(\dfrac{12.7 \text{ g HCl}}{1.00 \text{ L sol}} \right)\left(\dfrac{1 \text{ mol HCl}}{36.5 \text{ g HCl}} \right) = 0.348 \, M \text{ HCl}$

8.43 (a) 12 g of NaOH diluted to a 400.0 mL solution:

$$400.0 \text{ mL sol}\left(\frac{1 \text{ L sol}}{1000 \text{ mL sol}}\right)\left(\frac{0.75 \text{ mol NaOH}}{1 \text{ L sol}}\right)\left(\frac{40.0 \text{ g NaOH}}{1 \text{ mol NaOH}}\right) = 12 \text{ g NaOH}$$

(b) 12 g of $Ba(OH)_2$ diluted to a 1.0 L solution:

$$\frac{0.071 \text{ mol Ba(OH)}_2}{1 \text{ L sol}}\left(\frac{171.3 \text{ g Ba(OH)}_2}{1 \text{ mol Ba(OH)}_2}\right) = 12 \text{ g Ba(OH)}_2$$

(c) 3 g of KOH diluted to a 500.0 mL solution:

$$500.0 \text{ mL sol}\left(\frac{1 \text{ L sol}}{1000 \text{ mL sol}}\right)\left(\frac{0.1 \text{ mol KOH}}{1 \text{ L sol}}\right)\left(\frac{56.1 \text{ g KOH}}{1 \text{ mol KOH}}\right) = 3 \text{ g KOH}$$

(d) 5×10^1 g of sodium acetate diluted to 2.0 L of solution:

$$2.0 \text{ L sol}\left(\frac{0.3 \text{ mol NaC}_2\text{H}_3\text{O}_2}{1 \text{ L sol}}\right)\left(\frac{82.0 \text{ g NaC}_2\text{H}_3\text{O}_2}{1 \text{ mol NaC}_2\text{H}_3\text{O}_2}\right) = 5 \times 10^1 \text{ g NaC}_2\text{H}_3\text{O}_2$$

8.45 5.66 mL of 0.740 M H_2SO_4 are required to titrate 27.0 mL of 0.310 M NaOH.

$$27.0 \text{ mL NaOH sol}\left(\frac{1 \text{ L sol}}{1000 \text{ mL sol}}\right)\left(\frac{0.310 \text{ mol NaOH}}{1 \text{ L sol}}\right) = 8.37 \times 10^{-3} \text{ mol NaOH}$$

$$8.37 \times 10^{-3} \text{ mol NaOH}\left(\frac{1 \text{ mol H}_2\text{SO}_4}{2 \text{ mol NaOH}}\right)\left(\frac{1 \text{ L H}_2\text{SO}_4 \text{ sol}}{0.740 \text{ mol H}_2\text{SO}_4}\right)\left(\frac{1000 \text{ mL sol}}{1 \text{ L sol}}\right) = 5.66 \text{ mL}$$

8.47 Assuming that the base generates one mole of hydroxide per mole of base:

$$22.0 \text{ mL HCl sol}\left(\frac{0.150 \text{ mol HCl}}{1000 \text{ mL sol}}\right)\left(\frac{1 \text{ mol H}^+}{1 \text{ mol HCl}}\right) = 3.30 \times 10^{-3} \text{ mol H}^+$$

At the end point, 3.30×10^{-3} mol of H^+ added to 3.30×10^{-3} mol of the unknown base.

8.49 The point at which an indicator changes color is called the end point. It is usually so close to the equivalence point that difference between the two becomes insignificant.

8.51 In the CH_3COOH/CH_3COO^- buffer solution, the CH_3COO^- is completely ionized while the CH_3COOH is only partially ionized.

(a) $H_3O^+ + CH_3COO^- \rightleftharpoons CH_3COOH + H_2O$ (removal of H_3O^+)

(b) $HO^- + CH_3COOH \rightleftharpoons CH_3COO^- + H_2O$ (removal of OH^-)

8.53 Yes, the conjugate acid becomes the weak acid and the weak base becomes the conjugate base.

8.55 Altering the weak acid/conjugate base ratio according to the Henderson-Hasselbalch equation changes the buffer's pH. The buffer capacity can be changed without change in pH by increasing or decreasing the amount of weak acid/conjugate base mixture while keeping the ratio of the two constant.

8.57 This would occur in a couple of cases. One is very common, which is where you are using a buffer, such as TRIS with a pK_a of 8.3, but you do not want the solution to have a pH of 8.3. If you wanted a pH of 8.0, for example, you would have to have unequal amounts of the conjugate acid and base, with there being more conjugate acid. Another might be a situation where you are performing a reaction that you know will generate H^+ but you want the pH to be stable. In that situation, you might start with a buffer that was initially set to have more of the conjugate base so that it could absorb more of the H^+ that you know will be produced.

8.59 No. 100 mL of a 0.1 M phosphate buffer at pH 7.2 has a total of 0.01 moles of weak acid and 0.01 moles of the conjugate base. 20 mL of 1 M NaOH has 0.02 moles of base, so there is more total base than there is buffer to neutralize it. This buffer lacks the capacity to handle the amount of added NaOH.

8.61 (a) According to the Henderson-Hasselbalch equation, no change in pH will be observed as long as the ratio of weak acid/conjugate base remains the same.
      (b) The buffer capacity increases with increasing amount of weak acid/conjugate base. Therefore, 1.0 mol amounts of each diluted to 1 L would have a greater buffer capacity than 0.1 mol of each diluted to 1 L.

8.63 Using the Henderson-Hasselbalch Equation:

$$pH = pK_a + \log \frac{[A^-]}{[HA]} \qquad \text{When the } [A^-]:[HA] \text{ ratio is 10:1, the } \log \frac{[A^-]}{[HA]} = 1$$

and the equation reduces to: $pH = pK_a + 1$

8.65 When 0.10 mol of sodium acetate is added to 0.10 M HCl, the sodium acetate is completely neutralizes the HCl to acetic acid and sodium chloride. The pH of the solution is determined by the incomplete ionization of acetic acid.

$$K_a = \frac{[CH_3COO^-][H_3O^+]}{[CH_3COOH]} \qquad [H_3O^+] = [CH_3COO^-] = x$$

$$\sqrt{x^2} = \sqrt{K_a[CH_3COOH]} = \sqrt{(1.8 \times 10^{-5})(0.10)}$$

$$x = [H_3O^+] = 1.3 \times 10^{-3}\ M$$

$$pH = -log[H_3O^+] = 2.89$$

8.67 $[TRIS-H]^+ + NaOH \rightleftharpoons TRIS + H_2O$

8.69 The only parameter you need to know about a buffer is its pK_a. Choosing a buffer involves identifying the acid form that has a pK_a within one unit of the desired pH.

8.71 Choosing a buffer involves identifying the acid form that has a pK_a within one unit of the desired pH (a pH of 8.15). The TRIS buffer with a $pK_a = 8.3$ best fits this criteria.

8.73 $Mg(OH)_2$ is a weak base used as a flame-retardant in plastics.

8.75 (a) Respiratory acidosis is caused by hypoventilation, which is caused by a variety of breathing difficulties, such as a windpipe obstruction, asthma, or pneumonia.
(b) Metabolic acidosis is caused by starvation or heavy exercise.

8.77 Sodium bicarbonate is the weak base form of one of the blood buffers. It will tend to raise the pH of the blood, which is the purpose of the sprinter's trick, so that the person can absorb more H^+ during the event. By putting $NaHCO_3$ into the system, the following reaction will occur:

$$HCO_3^-(aq) + H^+(aq) \rightleftharpoons H_2CO_3(aq)$$

The loss of the H^+ means that the blood pH will rise.

8.79 The equilibrium favors the side of the weaker acid/weaker base.
(a) Benzoic acid is soluble in aqueous NaOH.

$$C_6H_5COOH + NaOH \rightleftharpoons C_6H_5COO^- + Na^+ + H_2O$$

$pK_a = 4.19$ $pK_a = 15.56$

(b) Benzoic acid is soluble in aqueous $NaHCO_3$.

$$C_6H_5COOH + NaHCO_3 \rightleftharpoons CH_3C_6H_4O^- + Na^+ + H_2CO_3$$

$pK_a = 4.19$ $pK_a = 6.37$

(c) Benzoic acid is soluble in aqueous Na_2CO_3.

$$C_6H_5COOH + Na_2CO_3 \rightleftharpoons CH_3C_6H_4O^- + Na^+ + HCO_3^-$$

$pK_a = 4.19$ $pK_a = 10.25$

8.81 The strength of an acid is not important to the amount of NaOH that would be required to hit a phenolphthalein endpoint. Therefore, the more concentrated acid, the acetic acid, would require more NaOH.

8.83 The solution of oxalic acid is 3.70×10^{-3} M.

$$\frac{0.583 \text{ g } H_2C_2O_4}{1.75 \text{ L sol}} \left(\frac{1 \text{ mol } H_2C_2O_4}{90.04 \text{ g } H_2C_2O_4} \right) = 3.70 \times 10^{-3} \text{ M oxalic acid}$$

8.85 The concentration of barbituric acid equilibrates to 0.90 M.

$$K_a = \frac{[\text{Barbiturate}^-][H_3O^+]}{[\text{Barbituric acid}]} \qquad x = [H_3O^+] = [\text{Barbiturate}^-]$$

$$[\text{Barbituric acid}] = \frac{x^2}{K_a} = \frac{(0.0030)^2}{1 \times 10^{-5}} = 0.9 \text{ M}$$

8.87 Yes, a pH = 0 is possible. A 1.0 M solution of HCl has a $[H_3O^+] = 1.0$ M.

pH = $-\log[H_3O^+] = -\log[1.0 \text{ M}] = 0.00$

8.89 The qualitative relationship between acids and their conjugate bases states that the stronger the acid, the weaker its conjugate base. This can be quantified in the equation: $K_b \times K_a = K_w$ or $K_b = 1.0 \times 10^{-14}/K_a$ where K_b is the base dissociation equilibrium constant for the conjugate base, K_a is the acid dissociation equilibrium constant for the acid, and K_w is the ionization equilibrium constant for water.

8.91 Yes. The strength of the acid is irrelevant. Both acetic acid and HCl have one H^+ to give up, so equal moles of either will require equal moles of NaOH to titrate to an endpoint.

8.93 Using the Henderson-Hasselbalch equation:

$$\frac{[H_2BO_3^-]}{[H_3BO_3]} = 10^{pH\text{-}pKa} = 10^{8.40\text{-}9.14}$$

$$\frac{[H_2BO_3^-]}{[H_3BO_3]} = 0.18$$

Need 0.18 mol of $H_2BO_3^-$ and 1.0 mol of H_3BO_3 in 1.0 L of solution.

8.95 Equilibria favor the side of the weaker acid/weaker base. Large pK_a values correlate with weak acids and small pK_a values correlate with strong acids, therefore, equilibria favor the side with the largest pK_a values.

8.97 (a) $HCOO^- + H_3O^+ \rightleftharpoons HCOOH + H_2O$

(b) $HCOOH + HO^- \rightleftharpoons HCOO^- + H_2O$

8.99 Using the Henderson-Hasselbalch Equation:

(a) $[Na_2HPO_4] = [NaH_2PO_4]10^{pH\text{-}pKa} = [0.050M]10^{7.21\text{-}7.21} = 0.050\ M$

(b) $[Na_2HPO_4] = [NaH_2PO_4]10^{pH\text{-}pKa} = [0.050M]10^{6.21\text{-}7.21} = 0.0050\ M$

(c) $[Na_2HPO_4] = [NaH_2PO_4]10^{pH\text{-}pKa} = [0.050M]10^{8.21\text{-}7.21} = 0.50\ M$

8.101 According to the Henderson-Hasselbalch equation:

$$pH = 7.21 + \log\frac{[HPO_4^{2-}]}{[H_2PO_4^-]}$$

As the concentration of $H_2PO_4^-$ increases, the $\log\frac{[HPO_4^{2-}]}{[H_2PO_4^-]}$ becomes negative, thus lowering the pH and becoming more acidic.

8.103 No. A buffer will only have a pH equal to its pK_a if there are equimolar amounts of the conjugate acid and base forms. If this is the basic form of TRIS, then just putting any amount of that into water will give a pH much higher than the pK_a value.

8.105 (a) pH = 7.10, $[H_3O^+]$ = 7.9 x 10^{-8} M, basic
(b) pH = 2.00, $[H_3O^+]$ = 1.0 x 10^{-2} M, acidic
(c) pH = 7.40, $[H_3O^+]$ = 4.0 x 10^{-8} M, basic
(d) pH = 7.00, $[H_3O^+]$ = 1.0 x 10^{-7} M, neutral
(e) pH = 6.60, $[H_3O^+]$ = 2.5 x 10^{-7} M, acidic
(f) pH = 7.40, $[H_3O^+]$ = 4.0 x 10^{-8} M, basic
(g) pH = 6.50, $[H_3O^+]$ = 3.2 x 10^{-7} M, acidic
(h) pH = 6.90, $[H_3O^+]$ = 1.3 x 10^{-7} M, acidic

8.107 Using the Henderson-Hasselbalch equation:

$$7.9 = 7.21 + \log\frac{[HPO_4^{2-}]}{[H_2PO_4^-]}$$

$$\frac{[HPO_4^{2-}]}{[H_2PO_4^-]} = 10^{0.69} = 4.9 \simeq 5 \text{ (to one significant figure)}$$

8.109 From a pH of 2.00, the $[H^+]$ = 10^{-pH} = 1.0 x 10^{-2} M

(a) $600. \text{ mL sol}\left(\dfrac{1 \text{ L sol}}{1000 \text{ mL sol}}\right)\left(\dfrac{1.0 \times 10^{-2} \text{ mol } H^+}{1 \text{ L sol}}\right) = 6.0 \times 10^{-3} \text{ mol } H^+ \text{ ions}$

(b) $6.0 \times 10^{-3} \text{ mol } H^+\left(\dfrac{1 \text{ mol NaHCO}_3}{1 \text{ mol } H^+}\right)\left(\dfrac{84.01 \text{ g NaHCO}_3}{1 \text{ mol NaHCO}_3}\right) = 0.50 \text{ g NaHCO}_3$

8.111 First we must determine how many moles of HF gas are dissolved in the 50 mL of water, assuming that all of the HF gas is dissolved into the water. This starts with unit conversions and using the ideal gas law equation: 20^oC = 293 K.

$$n = \frac{PV}{RT} = \frac{(0.601 \text{ atm})(1.00 \text{ L})}{(0.0821 \text{ L}\cdot\text{atm}\cdot\text{mol}^{-1}\cdot K^{-1})(293 \text{ K})} = 0.0250 \text{ mol HF}$$

(a) $\dfrac{0.0250 \text{ mol HF}}{50.0 \text{ mL sol}}\left(\dfrac{1000 \text{ mL sol}}{1 \text{ L sol}}\right) = 0.500 \text{ } M \text{ HF solution}$

(b) $[H^+]$ = 10^{-pH} = $10^{-1.88}$ = 0.013 M

$$HF(aq) \rightleftharpoons H^+(aq) + F^-(aq) \qquad K_a = \frac{[H^+][F^-]}{[HF]}$$

initial concentration   0.500 M 0 M 0 M

change   -x   +x   +x

equilibrium concentration   0.500 M-x   x   x

at equilibrium: x = $[H^+]$ = $[F^-]$ = 0.013 M

$$[HF] = 0.500\ M - x = 0.487\ M$$

$$K_a = \frac{[0.013\ M][0.013\ M]}{[0.487\ M]} = 3.5 \times 10^{-4}$$

8.113 At the equivalence point (neutralization, the number of OH$^-$ and H$^+$ ions reacted are equal. 1 mole of diprotic acid (dpa) reacts with 2 moles of OH$^-$.

$$30.0\ \text{mL NaOH sol}\left(\frac{1\ \text{L sol}}{1000\ \text{mL sol}}\right)\left(\frac{0.200\ \text{mol NaOH}}{1\ \text{L sol}}\right) = 6.00 \times 10^{-3}\ \text{mol NaOH}$$

$$6.00 \times 10^{-3}\ \text{mol NaOH}\left(\frac{1\ \text{mol dpa}}{2\ \text{mol NaOH}}\right) = 3.00 \times 10^{-3}\ \text{mol diprotic acid (dpa)}$$

$$\text{Molarity} = \frac{3.00 \times 10^{-3}\ \text{mol diprotic acid (dpa)}}{25.0\ \text{mL sol}}\left(\frac{1000\ \text{mL sol}}{1\ \text{L sol}}\right) = 0.120\ M\ \text{dpa}$$

$$\text{MW diprotic acid} = \frac{0.125\ \text{g dpa}}{3.00 \times 10^{-3}\ \text{mol}} = 41.7\ \text{g/mol}$$

Chapter 9: Nuclear Chemistry

9.1 $\quad {}^{139}_{53}I \rightarrow {}^{0}_{-1}e + {}^{139}_{54}Xe$

9.2 $\quad {}^{223}_{90}Th \rightarrow {}^{4}_{2}He + {}^{219}_{88}Ra$

9.3 $\quad {}^{74}_{33}As \rightarrow {}^{0}_{+1}e + {}^{74}_{32}Ge$

9.4 $\quad {}^{201}_{81}Th + {}^{0}_{-1}e \rightarrow {}^{201}_{80}Hg + \gamma$

9.5 Barium-122 (10 g) has decayed through 5 half-lives, leaving 0.31 g:
$10\,g \rightarrow 5.0\,g \rightarrow 2.5\,g \rightarrow 1.25\,g \rightarrow 0.625\,g \rightarrow 0.31\,g$
(careful consideration of significant figures ignored for clarity)

9.6 $\quad \text{Dose} = 50 \ \text{mCi}\left(\dfrac{9.0 \ \text{mL}}{300 \ \text{mCi}}\right) = 1.5 \ \text{mL}$

9.7 The intensity of any radiation decreases with the square of the distance: $\quad \dfrac{I_1}{I_2} = \dfrac{d_2^{\,2}}{d_1^{\,2}}$

assuming that 300 mCi is reported to one significant figure :

$$\frac{300. \ \text{mCi}}{I_2} = \frac{(3.0 \ \text{m})^2}{(0.010 \ \text{m})^2}$$

$$I_2 = \frac{(300. \ \text{mCi})(0.010 \ \text{m})^2}{(3.0 \ \text{m})^2} = 3.3 \times 10^{-3} \ \text{mCi}$$

9.9 Alpha particles are He^{2+} ions (${}^{4}_{2}He$) where protons are H^+ ions (${}^{1}_{1}H$).

9.11 $\lambda = \dfrac{c}{v}$

(a) $\lambda = \dfrac{3.0 \times 10^{10} \text{ cm/s}}{7.5 \times 10^{14} /s} = 4.0 \times 10^{-5} \text{ cm}$

$4.0 \times 10^{-5} \text{ cm} \left(\dfrac{1 \text{ m}}{100 \text{ cm}} \right) \left(\dfrac{10^9 \text{ nm}}{1 \text{ m}} \right) = 4.0 \times 10^2 \text{ nm}$ (visible light, blue)

(b) $\lambda = \dfrac{3.0 \times 10^{10} \text{ cm/s}}{1.0 \times 10^{10} /s} = 3.0 \text{ cm}$

$3.0 \text{ cm} \left(\dfrac{1 \text{ m}}{100 \text{ cm}} \right) \left(\dfrac{10^9 \text{ nm}}{1 \text{ m}} \right) = 3.0 \times 10^7 \text{ nm}$ (microwave)

(c) $\lambda = \dfrac{3.0 \times 10^{10} \text{ cm/s}}{1.1 \times 10^{15} /s} = 2.7 \times 10^{-5} \text{ cm}$

$2.7 \times 10^{-5} \text{ cm} \left(\dfrac{1 \text{ m}}{100 \text{ cm}} \right) \left(\dfrac{10^9 \text{ nm}}{1 \text{ m}} \right) = 2.7 \times 10^2 \text{ nm}$ (ultraviolet light)

(d) $\lambda = \dfrac{3.0 \times 10^{10} \text{ cm/s}}{1.5 \times 10^{18} /s} = 2.0 \times 10^{-8} \text{ cm}$

$2.0 \times 10^{-8} \text{ cm} \left(\dfrac{1 \text{ m}}{100 \text{ cm}} \right) \left(\dfrac{10^9 \text{ nm}}{1 \text{ m}} \right) = 2.0 \times 10^{-1} \text{ nm}$ (X-rays)

9.13 Longest wavelength: (a) infrared   Highest energy: (c) X-rays

9.15 For the lighter elements up to calcium, the stable isotopes have equal numbers of protons and neutrons. When this ratio unequal, the isotope is predicted to be radioactive.
(a) Nitrogen-13   (b) Phosphorus-33   (c) Lithium-9   (d) Calcium-39

9.17 For the lighter elements up to calcium, the stable isotopes have equal numbers of protons and neutrons. Oxygen-16 is more stable than the other two isotopes of oxygen because it has an equal number of protons and neutrons (eight of each).

9.19 $^{151}_{62}\text{Sm} \rightarrow ^{0}_{-1}\text{e} + ^{151}_{63}\text{Eu}$

9.21 $^{51}_{24}\text{Cr} + ^{0}_{-1}\text{e} \rightarrow ^{51}_{23}\text{V}$

9.23 $^{248}_{96}\text{Cm} + ^{28}_{10}\text{X} \rightarrow ^{116}_{51}\text{Sb} + ^{160}_{55}\text{Cs}$

The bombarding nucleus is $^{28}_{10}\text{Ne}$.

9.25 (a) $^{10}_{4}\text{Be} \rightarrow ^{0}_{-1}\text{e} + ^{10}_{5}\text{B}$ Beta emission

    (b) $^{151}_{63}\text{Eu}^m \rightarrow \gamma + ^{151}_{63}\text{Eu}$ Gamma emission

    (c) $^{195}_{81}\text{Tl} \rightarrow ^{0}_{+1}\text{e} + ^{195}_{80}\text{Hg}$ Positron emission

    (d) $^{239}_{94}\text{Pu} \rightarrow ^{4}_{2}\text{He} + ^{235}_{92}\text{U}$ Alpha emission

9.27 Gamma emission does not result in transmutation.

9.29 $^{239}_{94}\text{Pu} + ^{4}_{2}\text{He} \rightarrow ^{240}_{95}\text{Am} + ^{1}_{1}\text{H} + 2\,^{1}_{0}\text{n}$

9.31 Iodine-125 decayed through approximately six half-lives, with 0.31 mg remaining:
    20 mg $\rightarrow$ 10 mg $\rightarrow$ 5.0 mg $\rightarrow$ 2.5 mg $\rightarrow$ 1.25 mg $\rightarrow$ 0.625 mg $\rightarrow$ 0.31 mg
        (careful consideration of significant figures ignored for clarity)

9.33 The plutonium underwent four half-lives after the glacier deposited it.
1 mg $\rightarrow$ 2 mg $\rightarrow$ 4 mg $\rightarrow$ 8 mg $\rightarrow$ 16 mg:
Originally, there were 16 mg of plutonium/kg rock at the time of deposition.

9.35 The rate of radioactive decay is independent of all conditions and is a property of each specific isotope.

9.37 (a) Half-lives $= 2\text{ hrs}\left(\dfrac{1\text{ day}}{24\text{ hrs}}\right)\left(\dfrac{1\text{ half-life}}{8\text{ day}}\right) = 0.010$ half-lives after 2 hrs

Assuming that 200 mCi is reported to at least two significant figures:
Iodine-131 taken up by thyroid = 200 mCi x 0.12 = 24 mCi

Amount lost to decay (after 2 hrs) $= 24\text{ mCi}\left(\dfrac{0.010}{2}\right) = 0.12$ mCi (a negligible loss)

Iodine-131 remaining after 2 hrs = 24 mCi

$^{131}_{53}\text{I}$ remaining after 2 hrs $= 24\text{ mCi}\left(\dfrac{3.7 \times 10^7\text{ counts/s}}{1\text{ mCi}}\right) = 8.9 \times 10^8$ counts/s

    (b) After 24 days, three half-lives have passed; 1/2 x 1/2 x 1/2 = 1/8 or 12.5% of the original amount remains. 24.0 mCi x 0.125 = 3.0 mCi

9.39 Gamma radiation has the greatest penetrating power; therefore it requires the largest amount of shielding.

9.41 It would be best to stand at least 30 meters away if you wish to be subjected to no more than 0.20mCi.

$$\frac{I_1}{I_2} = \frac{d_2^2}{d_1^2}$$

$$d_2 = \sqrt{\frac{d_1^2 I_1}{I_2}} = \sqrt{\frac{(1.0 \text{ m})^2 (175 \text{ mCi})}{(0.20 \text{ mCi})}} = 3.0 \times 10^1 \text{ m}$$

9.43 A curie (Ci) measures radiation intensity.

9.45 $\text{Dose} = 7.2 \text{ mCi} \left(\frac{7.0 \text{ cc}}{80.0 \text{ mCi}} \right) = 0.63 \text{ cc}$

9.47 $\frac{I_1}{I_2} = \frac{d_2^2}{d_1^2}$

$$I_2 = \frac{d_1^2 I_1}{d_2^2} = \frac{(1 \text{ cm})^2 (1 \times 10^6 \text{ Bq})}{(20 \text{ cm})^2} = 3 \times 10^3 \text{ Bq}$$

$$I_2 = 3 \times 10^3 \text{ Bq} \left(\frac{1 \text{ disint/s}}{1 \text{ Bq}} \right) \left(\frac{1 \text{ Ci}}{3.7 \times 10^{10} \text{ disint/s}} \right) \left(\frac{10^6 \text{ μCi}}{1 \text{ Ci}} \right) = 8 \times 10^{-2} \text{ μCi}$$

9.49 Person A was exposed to the largest dose of radiation and injured more seriously.

$$\text{Person A exposure} = 3.0 \text{ Sv} \left(\frac{1 \text{ rem}}{0.01 \text{ Sv}} \right) \left(\frac{10^3 \text{ mrem}}{1 \text{ rem}} \right) = 3.0 \times 10^5 \text{ mrem}$$

Person B exposure = 0.50 mrem

9.51 Iodine-131 is concentrated in the thyroid; therefore it would be expected to induce thyroid cancer.

9.53 (a) Cobalt-60 is used for (4) cancer therapy.
(b) Thallium-201 is used in (1) heart scans and exercise stress tests.
(c) Tritium is used for (2) measuring water content of the body.
(d) Mercury-197 is used for (3) kidney scans.

9.55 $_1^2\text{H} + {}_1^3\text{H} \rightarrow {}_0^1\text{n} + {}_2^4\text{He} + \text{energy}$

9.57 $_{83}^{209}\text{Bi} + {}_{26}^{58}\text{Fe} \rightarrow {}_0^1\text{n} + {}_{109}^{266}\text{Mt}$

9.59 $_5^{10}\text{B} + {}_0^1\text{n} \rightarrow {}_2^4\text{He} + {}_3^7\text{Li}$

9.61 The assumption of a constant carbon-14 to carbon-12 ratio rests on two assumptions: (1) that the carbon-14 is continuously generated in the upper atmosphere by the production and decay of nitrogen-15 and (2) that carbon-14 is incorporated into carbon dioxide, CO_2, and other carbon containing compounds and then distributed worldwide as part of the carbon cycle. The continual formation of carbon-14; transfer of the isotope within the oceans, atmosphere, and biosphere; and the decay of living matter keep the supply of carbon-14 constant. The lifetime of a plant is short relative to the slow radioactive decay of carbon-14 to carbon-12; therefore the change in carbon-14 to carbon-12 ratio is negligible over the lifetime of a plant.

9.63 2009 − 1350 = 659 years (if the experiment was run in the year 2009)
659 years/5730 = 0.115 half-lives

9.65 Radon-222 produced polonium-218 by alpha emission:
$$_{86}^{222}\text{Rn} \rightarrow {}_2^4\text{He} + {}_{84}^{218}\text{Po}$$

9.66 Radioactive iodine-131 is concentrated in the thyroid where the radiation can induce thyroid cancer.

9.67 The decay of a radioactive isotope is an exponential curve:

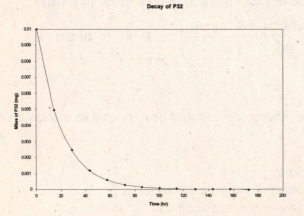

9.69 $^{19}_{10}\text{Ne} \rightarrow {}^{0}_{+1}\text{e} + {}^{19}_{9}\text{F}$

$^{20}_{11}\text{Na} \rightarrow {}^{0}_{+1}\text{e} + {}^{20}_{10}\text{Ne}$

9.71 Both the curie and the becquerel have units of disintegrations/second, a measurement of radiation intensity.

9.73 (a) $\dfrac{294 \text{ mrem/yr}}{359 \text{ mrem/yr}} \times 100\% = 81.9\%$

(b) $\dfrac{39 \text{ mrem/yr}}{359 \text{ mrem/yr}} \times 100\% = 11\%$

(c) $\dfrac{0.5 \text{ mrem/yr}}{359 \text{ mrem/yr}} \times 100\% = 0.1\%$

9.75 X-rays will cause more ionization than radar because X-rays have more energy.

9.77 1000/475 ~ 2.3 half-lives: 1/2 x 1/2 = 1/4 so a little less than 25% of the original americium will be around after 1000 years. Neptunium-237 is the decay product along with an alpha particle.

$^{241}_{95}\text{Am} \rightarrow {}^{4}_{2}\text{He} + {}^{237}_{93}\text{Np}$

9.79 One sievert is equal to 100 rem. This is a sufficient dose to cause radiation sickness but not certain death.

9.81 (a) Radioactive elements are constantly decaying to other isotopes and elements mixed in with the original isotopes.
(b) Beta emissions result from the decay of a neutron in the nucleus to a proton (the increase in atomic number) and an electron (beta particle).

9.83 Oxygen-16 is stable because it has an equal number of protons and neutrons. The others are unstable because the numbers of protons and neutrons are unequal. In this case, the greater the difference in numbers of protons and neutrons, the faster the isotope decays.

9.85 $^{208}_{82}\text{Pb} + {}^{64}_{28}\text{Ni} \rightarrow 6\,{}^{1}_{0}\text{n} + {}^{266}_{110}\text{X}$

The new element is Darmstadtium: $^{266}_{110}\text{Ds}$

9.87 Lithium-7 and He-4 are produced when boron control rods absorb neutrons.

$^{10}_{5}\text{B} + {}^{1}_{0}\text{n} \rightarrow {}^{11}_{5}\text{B}$

$^{11}_{5}\text{B} \rightarrow {}^{4}_{2}\text{He} + {}^{7}_{3}\text{Li}$

9.89 $^{222}_{86}\text{Rn}$ $\rightarrow$ $^{4}_{2}\text{He}$ + $^{218}_{84}\text{Po}$

$^{218}_{84}\text{Po}$ $\rightarrow$ $^{4}_{2}\text{He}$ + $^{214}_{82}\text{Pb}$

$^{214}_{82}\text{Pb}$ $\rightarrow$ $^{0}_{-1}\text{e}$ + $^{210}_{83}\text{Bi}$

$^{210}_{83}\text{Bi}$ $\rightarrow$ $^{4}_{2}\text{He}$ + $^{206}_{81}\text{Tl}$

$^{210}_{81}\text{Tl}$ $\rightarrow$ $^{0}_{-1}\text{e}$ + $^{210}_{82}\text{Pb}$

9.91 For any nuclear reaction, the mass numbers and the atomic numbers of the nuclei involved on both sides must balance. The first step to solving this problem involves first balancing the mass numbers by losing alpha particles:

$^{232}_{90}\text{Th}$ $\rightarrow$ $\rightarrow$ $\rightarrow$ $^{208}_{82}\text{Pb}$ The difference of 24 in the mass number can be balanced by adding six alpha partcles to the product side: 6 $^{4}_{2}\text{He}$.

$^{232}_{90}\text{Th}$ $\rightarrow$ $\rightarrow$ $\rightarrow$ $^{208}_{82}\text{Pb}$ + 6 $^{4}_{2}\text{He}$ + x $^{0}_{-1}\text{e}$

Then balance the atomic numbers through the loss of beta particles. This is accomplished by the loss of four beta particles (where x = 4)

$^{232}_{90}\text{Th}$ $\rightarrow$ $\rightarrow$ $\rightarrow$ $^{208}_{82}\text{Pb}$ + 6 $^{4}_{2}\text{He}$ + 4 $^{0}_{-1}\text{e}$

Now the equation is balanced on both sides with the atomic numbers adding up to 90 and mass numbers adding up to 232.

Chapter 10: Organic Chemistry

10.1 Following are Lewis structures showing all bond angles.

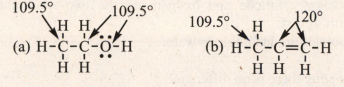

10.2 Of the four alcohols with molecular formula $C_4H_{10}O$, two are 1°, one is 2°, and one is 3°. For the Lewis structures of the 3° alcohol and two of the 1° alcohols, some $C-CH_3$ bonds are drawn longer to avoid crowding in the formulas.

$$H-\underset{\underset{H}{|}}{\overset{\overset{H}{|}}{C}}-\underset{\underset{H}{|}}{\overset{\overset{H}{|}}{C}}-\underset{\underset{H}{|}}{\overset{\overset{H}{|}}{C}}-\underset{\underset{H}{|}}{\overset{\overset{H}{|}}{C}}-\ddot{\underset{..}{O}}-H \qquad CH_3CH_2CH_2CH_2OH$$
Primary (1°)

$$H-\underset{\underset{H}{|}}{\overset{\overset{H}{|}}{C}}-\underset{|}{\overset{\overset{H-\overset{|}{C}-H}{|}}{C}}-\underset{\underset{H}{|}}{\overset{\overset{H}{|}}{C}}-\ddot{\underset{..}{O}}-H \qquad CH_3CHCH_2OH \overset{|}{\underset{CH_3}{}}$$
Primary (1°)

$$H-\underset{\underset{H}{|}}{\overset{\overset{H}{|}}{C}}-\underset{\underset{H}{|}}{\overset{\overset{:\ddot{O}:H}{|}}{C}}-\underset{\underset{H}{|}}{\overset{\overset{H}{|}}{C}}-\underset{\underset{H}{|}}{\overset{\overset{H}{|}}{C}}-H \qquad CH_3CH_2CHCH_3 \overset{OH}{\underset{|}{}}$$
Secondary (2°)

$$H-\underset{\underset{H-\overset{|}{C}-H}{\overset{|}{\underset{|}{H}}}}{\overset{\overset{H-\overset{|}{C}-H}{\overset{|}{H}}}{C}}-\underset{|}{\overset{}{C}}-\ddot{\underset{..}{O}}H \qquad CH_3\underset{\underset{CH_3}{|}}{\overset{\overset{CH_3}{|}}{C}}OH$$
Tertiary (3°)

10.3 The three secondary (2°) amines with the molecular formula $C_4H_{11}N$ are:

$$CH_3CH_2CH_2NHCH_3 \qquad CH_3\overset{\overset{CH_3}{|}}{C}HNHCH_3 \qquad CH_3CH_2NHCH_2CH_3$$

10.4 The three ketones with the molecular formula $C_5H_{10}O$ are:

$$CH_3CH_2CH_2\overset{\overset{O}{||}}{C}CH_3 \qquad CH_3CH_2\overset{\overset{O}{||}}{C}CH_2CH_3 \qquad CH_3\overset{\overset{O}{||}}{C}\underset{\underset{CH_3}{|}}{C}HCH_3$$

10.5 The two carboxylic acids with the molecular formula $C_4H_8O_2$ are:

$$CH_3CH_2CH_2\overset{\overset{O}{||}}{C}OH \qquad CH_3\underset{\underset{CH_3}{|}}{\overset{\overset{O}{||}}{C}}HCOH$$

10.6 The four esters with the molecular formula of $C_4H_8O_2$ are:

$$CH_3CH_2O\overset{\overset{O}{||}}{C}CH_3 \qquad CH_3CH_2\overset{\overset{O}{||}}{C}OCH_3 \qquad CH_3CH_2CH_2O\overset{\overset{O}{||}}{C}H \qquad CH_3\underset{\underset{CH_3}{|}}{C}HO\overset{\overset{O}{||}}{C}H$$

10. 7   (a) and (b): True
(c) False: carbon isn't even close. Silicon and oxygen are the two most abundant elements in the Earth's crust.
(d) False: most organic compounds are insoluble in water.

10.9   Assuming that each vanillin is pure, there is no difference.

10.11   Wöhler heated ammonium chloride and silver cyanate, both inorganic compounds, and obtained urea, an organic compound.

10.13   Carbon, which forms four bonds; hydrogen, which forms one bond; nitrogen, which forms three bonds and has one unshared pair of electrons; and oxygen, which forms two bonds and has two unshared pairs of electrons.

10.15   Following are Lewis dot structures for each element.

(a) ·C: (b) :Ö: (c) ·N: (d) :F:

10.17   Following are Lewis structures for each compound.

(a) H-Ö-Ö-H   (b) H-N̈-N̈-H   (c) H-C-Ö-H
                          H H

(d) H-C-S̈-H   (e) H-C-N̈-H   (f) H-C-Cl:

10.19   Following are Lewis structures for these ions. Under each ion is its name and the number of valence electrons it contains.

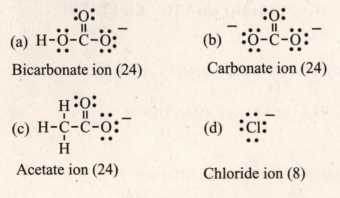

(a) H-Ö-C-Ö:⁻   (b) ⁻:Ö-C-Ö:⁻
Bicarbonate ion (24)   Carbonate ion (24)

(c) H-C-C-Ö:⁻   (d) :Cl:⁻
Acetate ion (24)   Chloride ion (8)

10.21 Each atom of C, O, and N is surrounded by regions of electron density. These regions of electron density are arranged in space so that there is the maximum separation between them. For four regions of electron density, the maximum separation occurs when they are arranged in a tetrahedral manner with angles of 109.5°. For three regions of electron density, the maximum separation occurs when they are arranged in a trigonal planar manner with angles of 120°. For two regions of electron density, the maximum separation occurs when they are arranged in a linear manner with a bond angle of 180°.

10.23 You would find two regions of electron density and, therefore, predict 180° for the C-O-H bond angle. The actual bond angle is approximately 109.5°.

10.25 (a) 120° about C and 109.5° about O   (b) 109.5° about N   (c) 120° about N

10.27 A functional group is a group of atoms in an organic molecule that undergoes a predictable set of chemical reactions.

10.29 The following functional groups are described by the following Lewis structures:

(a) $-\overset{\overset{\displaystyle :O:}{\|}}{C}-$ (b) $-\overset{\overset{\displaystyle :O:}{\|}}{C}-\overset{\cdot\cdot}{\underset{\cdot\cdot}{O}}-H$ (c) $-\overset{\cdot\cdot}{\underset{\cdot\cdot}{O}}-H$ (d) $-\overset{\overset{\displaystyle \cdot\cdot}{}}{\underset{\displaystyle H}{N}}-H$ (e) $-\overset{\overset{\displaystyle :O:}{\|}}{C}-\overset{\cdot\cdot}{\underset{\cdot\cdot}{O}}-\overset{|}{\underset{|}{C}}-$

10.31 When applied to alcohols, tertiary (3°) means that the carbon bearing the -OH group is bonded to three carbon atoms.

10.33 When applied to amines, tertiary (3°) means that the amine nitrogen is bonded to three carbon groups.

10.35 (a) The four primary (1°) alcohols with the molecular formula $C_5H_{12}O$ are:

$$CH_3CH_2CH_2CH_2CH_2OH \quad CH_3\underset{CH_3}{CH}CH_2CH_2OH \quad CH_3CH_2\underset{CH_3}{CH}CH_2OH \quad CH_3\overset{CH_3}{\underset{CH_3}{C}}CH_2OH$$

(b) The three secondary (2°) alcohols with the molecular formula $C_5H_{12}O$ are:

$$CH_3\overset{OH}{CH}CH_2CH_2CH_3 \quad CH_3CH_2\overset{OH}{CH}CH_2CH_3 \quad CH_3\overset{OH}{CH}\underset{CH_3}{CH}CH_3$$

(c) The one tertiary (3°) alcohol with the molecular formula $C_5H_{12}O$ is:

$$CH_3CH_2\overset{CH_3}{\underset{CH_3}{C}}OH$$

10.37 The eight carboxylic acids with the molecular formula $C_6H_{12}O_2$ are:

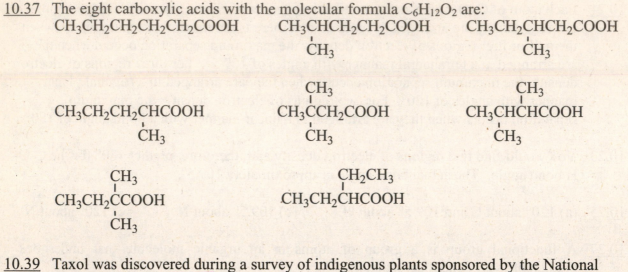

10.39 Taxol was discovered during a survey of indigenous plants sponsored by the National Cancer Institute with the goal of discovering new chemicals for fighting cancer. Taxol has shown anti-tumor activity for lung, breast, ovarian, and head/neck cancers.

10.41 Arrows point to atoms and show bond angles about each atom.

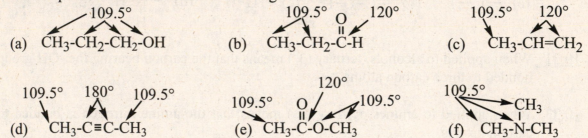

10.43 Predict 109.5° for the C-P-C bond angle.

10.45 The eight aldehydes with the molecular formula $C_6H_{12}O$ are:

$CH_3CH_2CH_2CH_2CH_2CH$ (O)

$CH_3CH_2CH_2CHCH$ (O), CH_3

$CH_3CH_2CHCH_2CH$ (O), CH_3

$CH_3CHCH_2CH_2CH$ (O), CH_3

$CH_3CHCHCH$ (O), CH_3, CH_3

CH_3CH_2C-CH (O), CH_3

CH_3CCH_2CH (O), CH_3, CH_3

CH_3CH_2CHCH (O), CH_2CH_3

10.47 As a general rule, if the difference in electronegativity between bonded atoms is less than 0.5 unit, the bond is nonpolar covalent; if it is between 0.5 and 1.9, the bond is polar covalent.
(a) C-C (2.5 - 2.5 = 0.0) nonpolar covalent
(b) C=C (2.5 - 2.5 = 0.0) nonpolar covalent
(c) C-H (2.5 - 2.1 = 0.4) nonpolar covalent
(d) C-O (3.5 - 2.5 = 1.0) polar covalent
(e) O-H (3.5 - 2.1 = 1.4) polar covalent
(f) C-N (3.0 - 2.5 = 0.5) polar covalent
(g) N-H (3.0 - 2.1 = 0.9) polar covalent
(h) N-O (3.5 - 3.0 = 0.5) polar covalent

10.49 Under each formula is given the difference in electronegativity between the atoms of the most polar bond.

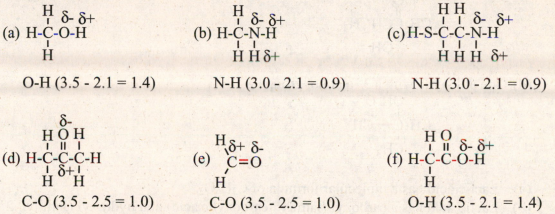

(a) O-H (3.5 - 2.1 = 1.4)   (b) N-H (3.0 - 2.1 = 0.9)   (c) N-H (3.0 - 2.1 = 0.9)

(d) C-O (3.5 - 2.5 = 1.0)   (e) C-O (3.5 - 2.5 = 1.0)   (f) O-H (3.5 - 2.1 = 1.4)

10.51 (a) All of the carbons in naphthalene have three areas of electron density. The VSEPR theory predicts that each carbon atom is trigonal planar. All of the carbon atoms in naphthalene form a planar structure.

(b) Naphthalene is composed of only C-C and C-H bonds, neither of which is polar. Therefore, the molecule is nonpolar.

10.53 The following compounds have a molecular formula of $C_4H_8O_2$:
(a) Two carboxylic acids:

$$CH_3CH_2CH_2\overset{\overset{\displaystyle O}{\|}}{C}OH \qquad CH_3\underset{\underset{\displaystyle CH_3}{|}}{\overset{\overset{\displaystyle O}{\|}}{C}}HCOH$$

(b) Four esters:

$$CH_3CH_2O\overset{\overset{\displaystyle O}{\|}}{C}CH_3 \quad CH_3CH_2\overset{\overset{\displaystyle O}{\|}}{C}OCH_3 \quad CH_3CH_2CH_2O\overset{\overset{\displaystyle O}{\|}}{C}H \quad CH_3\underset{\underset{\displaystyle CH_3}{|}}{C}HO\overset{\overset{\displaystyle O}{\|}}{C}H$$

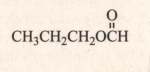

(c) One ketone with a 2° alcohol group:

$$CH_3\overset{\overset{\displaystyle O}{\|}}{C}\underset{\underset{\displaystyle OH}{|}}{C}HCH_3$$

(d) One aldehyde with a 3° alcohol group:

$$CH_3\underset{\underset{\displaystyle CH_3}{|}}{\overset{\overset{\displaystyle OH}{|}}{C}}\overset{\overset{\displaystyle O}{\|}}{-CH}$$

10.55 (a) Lactic acid has a molecular formula of $C_3H_6O_3$.
(b) Lactic acid has a carboxyl (family: carboxylic acid) and hydroxyl (family: 2° alcohol) functional groups.
(c) All bond angles around the indicated atoms are predicted using the VSEPR theory.

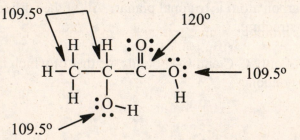

(d) The polar bonds include the C-O, C=O, and the two O-H bonds.
(e) Lactic acid has several polar bonds, the C-O, C=O, and the O-H being the most polar. The geometry of the molecule allows a combination of polar bonds to result in a polar molecule.

10.57 The theoretical yield for the reaction between 120 g of salicylic acid (SA) and excess acetic anhydride is 1.6×10^2 g of aspirin (ASP).

$$\text{theor. yield} = 120 \text{ g SA} \left(\frac{1 \text{ mol SA}}{138.1 \text{ g SA}} \right) \left(\frac{1 \text{ mol ASP}}{1 \text{ mol SA}} \right) \left(\frac{180.2 \text{ g ASP}}{1 \text{ mol ASP}} \right) = 1.6 \times 10^2 \text{ g}$$

10.59 (a) Two contributing structures for acetamide:

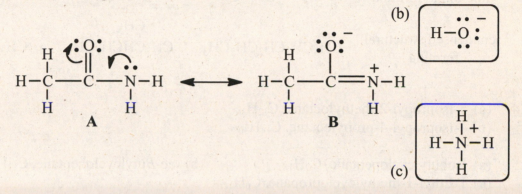

(b) The oxygen atom in hydroxide has only one bond, three unshared pairs of electrons, and a negative charge, which is similar to the oxygen atom in structure **B**.

(c) The nitrogen atom in ammonium has four bonds and a positive charge, as does the nitrogen atom in structure **B**.

(d) If structure **A** is a major contributor to the hybrid structure, the VSEPR theory predicts $109°$ bond angles around the nitrogen to best minimize the four electron pair repulsions.

(e) If structure **B** is a major contributor to the hybrid structure, the VSEPR theory predicts $120°$ bond angles around the nitrogen to best minimize the repulsions imposed by three regions of electron density.

(f) The discovery of $120°$ bond angles around the nitrogen in protein amide bonds suggests that structure **B** is the more important contributing structure in the resonance hybrids.

11.1 The alkane is octane, and its molecular formula is C_8H_{18}.

11.2 (a) Constitutional isomers   (b) Same compound

11.3 The three constitutional isomers with the molecular formula C_5H_{12} are:

line-angle formula

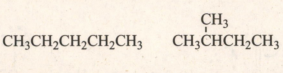

condensed structural formula $CH_3CH_2CH_2CH_2CH_3$ $CH_3CHCH_2CH_3$ CH_3CCH_3

11.4 (a) 5-Isopropyl-2-methyloctane, $C_{12}H_{26}$
(b) 4-Isopropyl-4-propyloctane, $C_{14}H_{30}$

11.5 (a) Isobutylcyclopentane, C_9H_{18} (b) *sec*-Butylcycloheptane, $C_{11}H_{22}$
(c) 1-Ethyl-1-methylcyclopropane, C_6H_{12}

11.6 The structure with three methyl groups equatorial is:

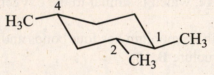

11.7 Cycloalkanes (a) and (c) show *cis-trans* isomerism.

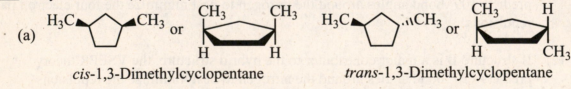

cis-1,3-Dimethylcyclopentane   *trans*-1,3-Dimethylcyclopentane

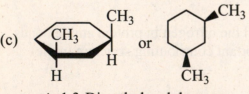   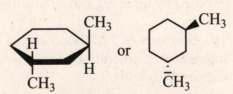

cis-1,3-Dimethylcyclohexane   *trans*-1,3-Dimethylcyclohexane

11.8 Boiling points of an isomeric series of alkanes increase with a decreasing degree of branching. In order of increasing boiling point, they are:
(a) 2,2-dimethylpropane (9.5°C), 2-methylbutane (27.8°C), pentane (36.1°C)
(b) 2,2,4-trimethylhexane (127 °C), 3,3-dimethylheptane (137 °C), nonane (151 °C)

11.9   The two products with the formula C_3H_7Cl are:

                                                       Cl
                                                       |
                    $CH_3CH_2CH_2Cl$ CH_3CHCH_3
                    1-Chloropropane   2-Chloropropane
                    (Propyl chloride)   (Isopropyl chloride)

11.11   (a) A hydrocarbon is a compound that contains only carbon and hydrogen.
        (b) An alkane is a saturated hydrocarbon.
        (c) A saturated hydrocarbon contains only carbon-carbon single bonds.

11.13   In a line-angle formula, carbon atoms are represented by the intersections of lines and a
        line terminus. Combinations of one, two, or three lines represent single, double, or triple
        bonds between atoms.

11.15   (a) $C_{10}H_{22}$ (b) C_8H_{18} (c) $C_{11}H_{24}$

11.17   Statements (a) and (b) are true. Statements (c) and (d) are false.

11.19   Structures (a) and (g) represent the same compound. Structures (a, g), (c), (d), (e), and (f)
        represent constitutional isomers with molecular formula $C_4H_{11}N$.

11.21   The following pairs are constitutional isomers:
        (b) both have the formula C_5H_8
        (c) both have the formula C_4H_8O
        (e) both have the formula $C_5H_{11}N$
        (f) both have the formula C_4H_8O

11.23   (a), (b), and (c): True

11.25   2-Methylpropane and 2-Methylbutane

11.27   (a) True
        (b) False: hexane has a molecular formula of C_6H_{14} and cyclohexane has a molecular
            formula of C_6H_{12}.
        (c) False: The parent name of a cycloalkane is the name of the unbranched linear alkane
            with the same number of carbon atoms that are in the ring, with the prefix cyclo.

11.29 (a)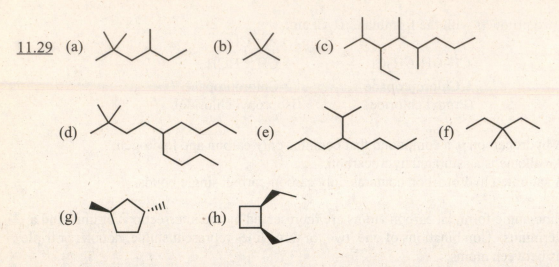

11.31 A condensed structural formula only shows the order of bonding of the atoms in a compound. There is no attempt in it to show bond angles and molecular shape.

11.33 The formula for calculating the interior angles of a regular polygon is:

$$\text{interior angle} = \frac{(n-2)180^\circ}{n} \quad \text{where } n = \text{number of sides}$$

(a) Cyclopropane is an equilateral triangle, so the C-C-C bond angles are calculated to be 60°. The optimal bond angle of a tetrahedral carbon is 109.5°. This drastic difference between calculated and optimal is the source of large angle strain and is the major contributor to cyclopropane's ring strain.

(b) Cyclopentane is a regular pentagon and the C-C-C bond angles are calculated to be 108°. The optimal bond angle of a tetrahedral carbon is 109.5°. The strain due to the difference between optimal and calculated is minimal.

11.35 Restricted rotation about the carbon-carbon single bonds of the ring makes *cis-trans* isomers possible.

11.37 The *cis* and *trans* isomers of 1,2-dimethylcyclopropane are drawn here in two ways. In the left drawing for each isomer, the ring is in the plane of the paper and methyl groups are above or below the plane. In the right drawing for each isomer, the ring is perpendicular to the plane of the paper and methyl groups are in the plane of the paper.

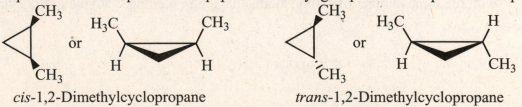

cis-1,2-Dimethylcyclopropane   *trans*-1,2-Dimethylcyclopropane

11.39 When equatorial, a group is as far away as possible from other atoms of the cyclohexane ring. When axial, it comes very close to two axial hydrogen atoms on the same side of the ring.

11.41 (a) Only one isomer exists for a cyclohexane with both a hydroxyl and a methyl substituent that does not show *cis/trans* isomerism:

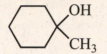

1-methylcyclohexanol

(b) There are three constitutions for methylcyclohexanol where *cis/trans* isomerism exists: 2-methylcyclohexanol, 3-methylcyclohexanol, and 4-methylcyclohexanol.

    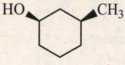   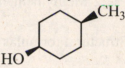

cis-2-methylcyclohexanol   *cis*-3-methylcyclohexanol   *cis*-4-methylcyclohexanol

    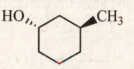   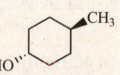

trans-2-methylcyclohexanol   *trans*-3-methylcyclohexanol   *trans*-4-methylcyclohexanol

11.43 The isomer with the highest boiling point is heptane (98°C), which is unbranched. The isomer with the lowest boiling point is 2,2-dimethylpentane (79°C), one of the most branched isomers.

11.45 Alkanes are less dense than water. Those that are liquid at room temperature float on water.

11.47 Both hexane and octane are colorless liquids and you cannot tell the difference between them by looking at them. One way to distinguish the difference between them is to determine their boiling points. Hexane has the lower molecular weight, thus the lower boiling point (69°C). Octane has the higher molecular weight and the higher boiling point (126°C).

11.49 Boiling point increases as the molecular weight increases.

11.51 Balanced equations for the complete combustion of each are:

(a) $2CH_3(CH_2)_4CH_3 + 19O_2 \longrightarrow 12CO_2 + 14H_2O$
Hexane

(b) ⬡ $+ 9O_2 \longrightarrow 6CO_2 + 6H_2O$
Cyclohexane

(c) $2CH_3\overset{CH_3}{\underset{|}{C}}HCH_2CH_3CH_3 + 19O_2 \longrightarrow 12CO_2 + 14H_2O$
2-Methylpentane

11.53 Structural formulas for each part are:

(a) CH_3Br (b) ⬡—Cl (c) $BrCH_2CH_2Br$ (d) $CH_3\overset{CH_3}{\underset{Cl}{C}}CH_3$ (e) CCl_2F_2

11.55 Both hexane and 2-methylpentane have the molecular formula of C_6H_{14}.

$$2C_6H_{14} + 19O_2 \rightarrow 12CO_2 + 14H_2O$$

(a) $2CH_3(CH_2)_4CH_3 + 19O_2 \longrightarrow 12CO_2 + 14H_2O$
Hexane

(b) ⬡ $+ 9O_2 \longrightarrow 6CO_2 + 6H_2O$
Cyclohexane

(c) $2CH_3\overset{CH_3}{\underset{|}{C}}HCH_2CH_3CH_3 + 19O_2 \longrightarrow 12CO_2 + 14H_2O$
2-Methylpentane

11.57

isomers of
C_5H_{12}

monochlorination product(s)

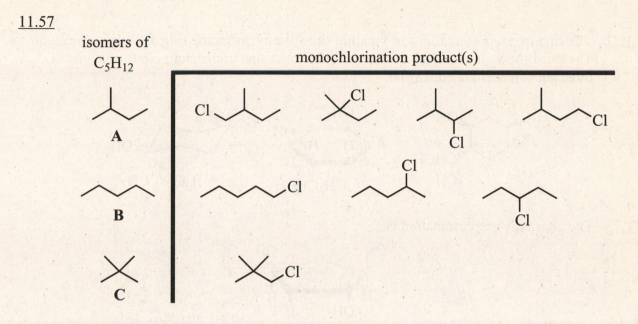

11.59 (a) Only one ring contains just carbon atoms.
(b) One ring contains two nitrogen atoms.
(c) One ring contains two oxygen atoms.

11.61 Octane will produce more engine knocking than heptane.

11.63 (a) The Freons are a class of chlorofluorocarbons.
(b) They were ideal for use as heat transfer agents in refrigeration systems because they are nontoxic, not corrosive, and nonflammable.
(c) Two Freons used for this purpose were CCl_3F (Freon-11) and CCl_2F_2 (Freon-12).

11.65 They are hydrofluorocarbons and hydrochlorofluorocarbons. These compounds are much more chemically reactive in the atmosphere than the Freons and are destroyed before they reach the stratosphere.

11.67 (a) The longest chain is pentane. Its IUPAC name is 2-methylpentane.
(b) The pentane chain is numbered incorrectly. Its IUPAC name is 2-methylpentane.
(c) The longest chain is pentane. Its IUPAC name is 3-ethyl-3-methylpentane.
(d) The longest chain is hexane. Its IUPAC name is 3,4-dimethylhexane.
(e) The longest chain is heptane. Its IUPAC name is 4-methylheptane.
(f) The longest chain is octane. Its IUPAC name is 3-ethyl-3-methyloctane.
(g) The ring is numbered incorrectly. Its IUPAC name is 1,1-dimethylcyclopropane.
(h) The ring is numbered incorrectly. Its IUPAC name is 1-ethyl-3-methylcyclohexane.

11.69 Tetradecane is a liquid at room temperature.

11.71 The first two representations of menthol show the cyclohexane ring as a planar hexagon. The third shows it as a chair conformation. Notice that in the chair conformation, all three substituents are equatorial.

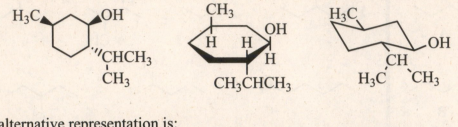

11.73 The alternative representation is:

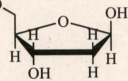

11.75 (a) Dimethyl ether has the lower boiling point (-24°C) than ethanol (78°C). Ethanol engages in strong intermolecular attractive forces through hydrogen bonding, therefore requires more energy to break them than dimethyl ether, which requires less energy to break weaker dipole-dipole and London dispersion forces to boil.

(b) Ethanol has the infinite water solubility because it participates in hydrogen bonding with water, acting as both hydrogen bond donor and acceptor. Diethyl ether is less soluble in water because it can only act as a hydrogen bond acceptor with water.

Chapter 12: Alkenes and Alkynes

12.1 (a) 3,3-Dimethyl-1-pentene   (b) 2,3-Dimethyl-2-butene   (c) 3,3-Dimethyl-1-butyne

12.2 (a) *trans*-3,4-Dimethyl-2-pentene   (b) *cis*-4-Ethyl-3-heptene

12.3 (a) 1-Isopropyl-4-methylcyclohexene   (b) Cyclooctene   (c) 4-*tert*-Butylcyclohexene

12.4 Line-angle formulas for the other two dienes are:

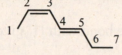 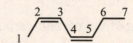

cis,trans-2,4-Heptadiene   *cis,cis*-2,4-Heptadiene

12.5 Four *cis-trans* isomers are possible.

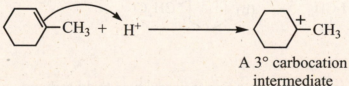

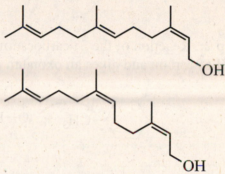

12.6 (a) CH₃CHCH₃ (Br)   (b) (cyclohexane with Br and CH₃)

12.7 Propose a two-step mechanism similar to that for the addition of HCl to propene.

Step 1: Reaction of H⁺ with the carbon-carbon double bond gives a 3° carbocation intermediate.

(reaction scheme) —CH₃ + H⁺ ⟶ ⁺—CH₃

A 3° carbocation
intermediate

Step 2: Reaction of the 3° carbocation intermediate with bromide ion completes the valence shell of carbon and gives the product.

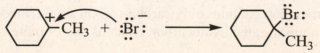

(reaction scheme) ⁺—CH₃ + :Br:⁻ ⟶ Br:—CH₃

107

12.8   The product from each acid-catalyzed hydration is the same alcohol.

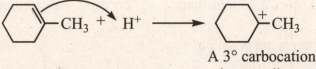

$$CH_3CCH_2CH_3$$

with CH_3 above and OH below the central carbon.

12.9   Propose a three-step mechanism similar to that for the acid-catalyzed hydration of propene.

Step 1: Reaction of the carbon-carbon double bond with H$^+$ gives a 3° carbocation intermediate.

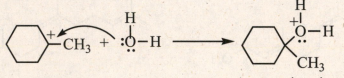

A 3° carbocation
intermediate

Step 2: Reaction of the 3° carbocation intermediate with water completes the valence shell of carbon and gives an oxonium ion.

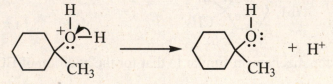

An oxonium ion

Step 3: Loss of H$^+$ from the oxonium ion completes the reaction and generates a new H$^+$ catalyst.

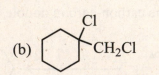

$+$ H$^+$

12.10   (a) CH$_3$-C-CH-CH$_2$ with CH$_3$ above, and H$_3$C, Br, Br below   (b) cyclohexane ring with Cl and CH$_2$Cl

12.11   (d) True
        (a) False: arenes are also classified as unsaturated hydrocarbons.
        (b) False: most of the ethylene comes from the cracking of petroleum.
        (c) False: ethylene, C$_2$H$_4$ and acetylene C$_2$H$_2$ are different compounds.

12.13   A saturated hydrocarbon contains only carbon-carbon single bonds and carbon-hydrogen single bonds. An unsaturated hydrocarbon contains one or more carbon-carbon double or triple bonds.

12.15 Start with a pentene parent chain and move the double bond down the chain, considering the stereoisomers, then proceed the same way with a butene parent chain. There are no possibilities with a propene parent chain.

pentene parent chain:

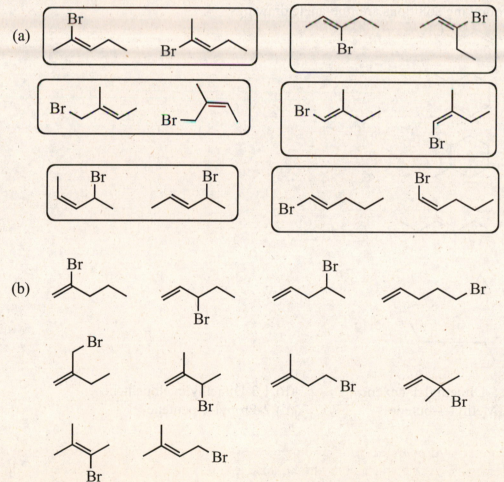

1-Pentene   *cis*-2-Pentene   *trans*-2-Pentene

butene parent chain:

2-Methyl-1-butene   2-Methyl-2-butene   3-Methyl-1-butene

12.17 Part (a) lists *cis/trans* isomer possibilities. Part (b) includes isomers that do not show *cis/trans* isomerism.

(a)

(b)

12.19 (a)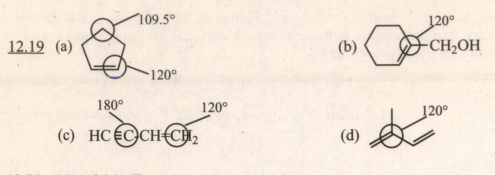

(b) 120° ⟨hexene⟩—CH₂OH

(c) HC≡C-CH=CH₂   180°   120°

(d) 120°

12.21 (a) and (e): True
(b) False: the IUPAC name is 2-butene.
(c) False: there are two of the same substituent on one of the carbons of the double bond, therefore there cannot be *cis/trans* isomerism.
(d) False: The four carbon-carbon bonds of the double bond are all in the same plane.
(f) False: neither of the double bonds in 1,3-butadiene are stereocenters.

12.23 The following solutions are line-angle drawings:

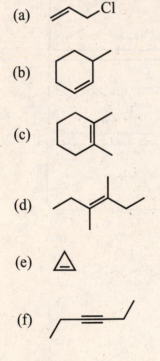

(a) ⟍⟋⟍Cl

(b)

(c)

(d)

(e) △

(f)

12.25 (a) 2,5-Dimethyl-1-hexene   (b) 1,3-Dimethylcyclopentene
(c) 2-Methyl-1-butene   (d) 2-Propyl-1-pentene

12.27 (a) The longest chain is a butene. The correct name is 2-methyl-1-butene.
(b) The ring is not numbered correctly. The correct name is 4-isopropylcyclohexene.
(c) The parent chain is not numbered correctly. The correct name is 3-methyl-2-hexene.
(d) The parent chain is five carbons. The correct name is 2-ethyl-3-methyl-1-pentene.
(e) The ring is not numbered correctly. The correct name is 3,3-dimethylcyclohexene
(f) The parent chain is seven carbons. The correct name is 3-methyl-3-heptene.

12.29 Parts (b), (c), and (e) show *cis-trans* isomerism. The following structures are line-angle formulas for the *cis* and *trans* isomers of each.

(b)   (c)

   cis-2-Hexene   *trans*-2-Hexene   *cis*-3-Hexene   *trans*-3-Hexene

(e)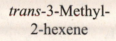

   cis-3-Methyl-   *trans*-3-Methyl-
   2-hexene   2-hexene

12.31 *Trans* double bonds are too unstable to exist in cycloalkenes with less than eight atoms.

   cis-cyclodecene   *trans*-cyclodecene

12.33 The line-angle drawing of oleic acid and elaidic acid:
Elaidic acid: COOH

Oleic acid: COOH

12.35 Only compounds (b) and (d) show cis-trans isomerism.

(b)

(d)

12.37 The structural formula of β-ocimene is drawn on the left showing all atoms and on the right as a line-angle formula.

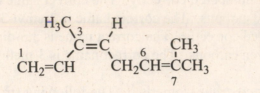

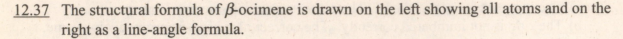

12.39 (a), (b), (c), (d), (f), (g), (h), (i), and (j): True
    (e) False: carbocations only have three bonds.
    (k) False: the product will be cyclohexane.
    (l)   False: the hydrogen comes from the catalyst and the hydroxyl comes from water.
    (m) False: it is a hydration (addition) reaction.
    (n) False: the hydration of 1-butene gives 2-butanol.

12.41 (a) HBr   (b) H_2O/H_2SO_4 (c) HI   (d) Br_2

12.43 Carbocation structures:

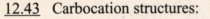

$$\underset{\text{Tertiary}}{CH_3CH_2\overset{+}{\underset{|}{C}}-CH_2CH_3} \quad \text{and} \quad \underset{\text{Secondary}}{CH_3CH_2\overset{\overset{CH_3}{|}}{C}H-\overset{+}{C}HCH_3}$$

(a)

(b) $\underset{\text{Secondary}}{CH_3CH_2\overset{+}{C}H-CH_2CH_3}$ and $\underset{\text{Secondary}}{CH_3CH_2CH_2-\overset{+}{C}HCH_3}$

 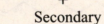 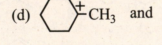

(c)   Tertiary   and   Secondary   (d) Tertiary   and   Primary

12.45 (a) 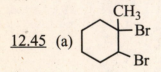   (b)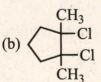

12.47 (a) $CH_3\overset{\overset{CH_3}{|}}{C}=CHCH_3$ (b) $CH_2=\overset{\overset{CH_3}{|}}{C}CH_2CH_3$ (c) $CH_2=CHCH_2CH_2CH_3$

12.49 The acid catalyzed hydration of alkenes follows the Markovnikov rule for regioselectivity, meaning that when unsymmetrical alkenes react with H_2O/H_3O^+, the carbon of the C=C that has the most hydrogen atoms, gets protonated. The carbon of the C=C that has the least number of hydrogen atoms, gets the hydroxyl from water.

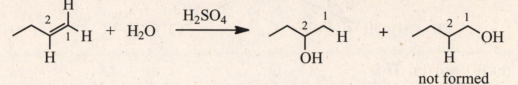

Carbon #1 from the alkene has two protons and carbon #2 has only one proton. Consequently, carbon #1 will get protonated and carbon #2 will get the hydroxyl, thus forming the secondary alcohol, 2-butanol, as the major product with very little of the primary alcohol 1-butanol formed.

12.51 2-Pentene is an unsymmetrical alkene meaning that the two groups attached to the alkene carbons are different. Also, both C2 and C3 of the double bond have one attached hydrogen atom, therefore, there will be no Markovnikov rule selectivity between C2 or C3 for which one adds the new proton and which one gets the hydroxyl. The result of the lack of selectivity means that the hydration of 2-pentene results in two products, 2-pentanol and 3-pentanol in roughly a 1:1 ratio.

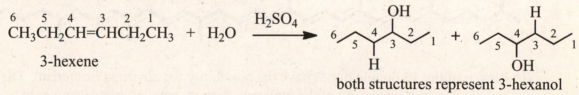

3-Hexene is a symmetrical alkene; meaning carbons C3 and C4 are identical and will protonate with equal probability. Consequently, it doesn't matter which carbon, C3 or C4, adds the new proton and which one gets the hydroxyl. The result of the hydration of 3-hexene is only one alcohol, 3-hexanol.

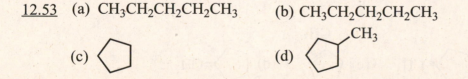

12.53 (a) $CH_3CH_2CH_2CH_2CH_3$ (b) $CH_3CH_2CH_2CH_2CH_3$

(c) ⬠   (d) ⬠CH_3

12.55 Reagents are shown over each arrow.

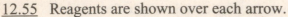

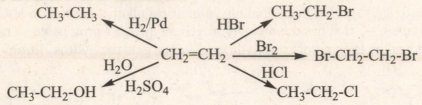

12.57 (c), (d), and (e): True
  (a) False: polyethylene has no carbon-carbon double bonds. Most polymers are named after their monomers, not their structural features.
  (b) False: all polyethylenes have C-C-C bond angles of approximately 109.5°.
  (f) False: only HDPE and PET are recycled in large quantities.

12.59 Pheromones are chemicals used by the animal world to communicate.

12.61 1×10^{-12} g TDA $\left(\dfrac{1 \text{ mol TDA}}{254.4 \text{ g TDA}}\right)\left(\dfrac{6.02 \times 10^{23} \text{ molecules TDA}}{1 \text{ mol 11-TDA}}\right) = 2 \times 10^9$ molec TDA

12.63 The all *trans*-retinal has the longer end-to-end distance compared to 11-*cis*-retinal. The *cis* double bond in 11-*cis*-retinal imparts a bend in the molecule.

12.65 (a) V: polyvinyl chloride
  (b) PP: polypropylene
  (c) PS: polystyrene

12.67 (a) The carbon skeleton of lycopene can be divided into eight isoprene units, here shown in bold bonds.

  (b) Eleven of the 13 double bonds have the possibility for *cis-trans* isomerism. The double bonds at either end of the molecule cannot show *cis-trans* isomerism.

12.69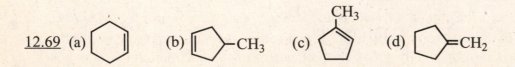

12.71 The alcohol is 3-hexanol. Each alkene gives the same 2° carbocation intermediate and the same alcohol.

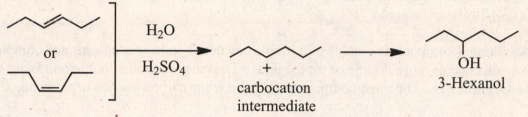

12.73 Reagents are shown over the arrows.

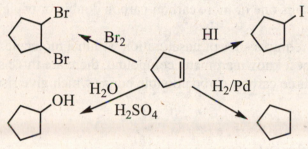

12.75 Line-angle formulas for these three unsaturated fatty acids are

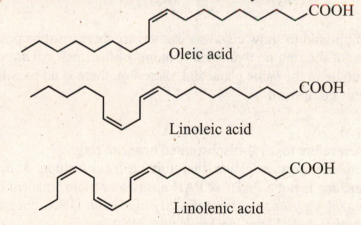

12.77 Reaction (1) is a dehydrogenation (elimination of hydrogen), reaction (2) is a hydration (addition of water), and reaction (3) is an oxidation. The reaction sequence is called a β-oxidation because the β-carbon to the carbonyl is oxidized. It is common in organic and biochemistry to call a carbon adjacent to a functional group an α-carbon and the next one a β-carbon. In this biochemical sequence, the β-carbon from the carbonyl group of the ester is converted from a -CH₂- group to a ketone group by an oxidation reaction, hence the term β-oxidation.

Chapter 13: Benzene and Its Derivatives

13.1   (a) 2,4,6-Tri-*tert*-butylphenol   (b) 2,4-Dichloroaniline
        (c) 3-Nitrobenzoic acid

13.3   A saturated compound contains only single covalent bonds. An unsaturated compound contains one or more double or triple bonds. The most common double bonds are C=C, C=O, and C=N. The most common triple bond is the carbon-carbon triple bond, $C \equiv C$.

13.5   The members of each class of hydrocarbon contain fewer hydrogen atoms than an alkane or cycloalkane with the same number of carbons. Alternatively, each class of hydrocarbon contains one or more carbon-carbon double or triple bonds.

13.7   No. Aromaticity requires sites of unsaturation within a molecule. In the case of benzene, the simplest and best known aromatic compound, there are three sites of unsaturation; that is, there are three carbon-carbon double bonds which give rise to the so-called aromatic sextet.

13.9   (a) CH_4 (b) $CH_2=CH_2$ (c) $HC \equiv CH$ (d)

        Methane   Ethene   Ethyne   Benzene
                     (Ethylene)   (Acetylene)

13.11   For a cyclic compound to show *cis-trans* isomerism, there must be positions above and below the plane of the ring on two or more atoms of the ring. All atoms of 1,4-dichlorobenzene lie in the same plane and, therefore, there is no possibility for *cis-trans* isomerism in this compound.

13.13   (a) and (d): True
        (b)   False: *para* refers to a 1,4-disubstituted benzene ring.
        (c)   False: benzene rings are planar, therefore *cis/trans* isomers do not exist.
        (e)   False: benzene is not a PAH. A PAH has two or more adjacent benzene rings.
        (f)   False: benzo[A]pyrene does not directly bind with DNA, but instead, its metabolic product, a diol epoxide, binds with DNA and causes mutations.

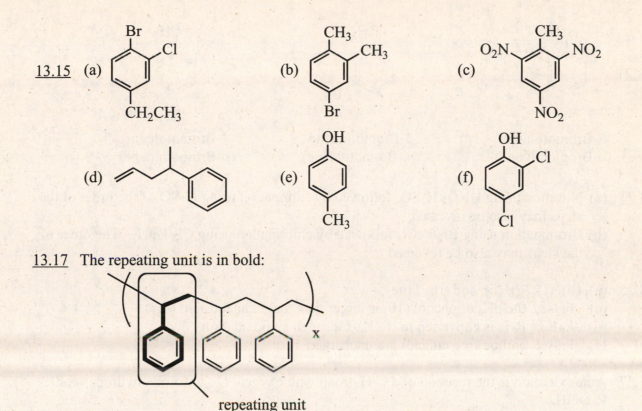

13.15

13.17 The repeating unit is in bold:

repeating unit

13.19 Only cyclohexene will react with a solution of bromine in dichloromethane. A solution of Br_2/CH_2Cl_2 is red, whereas a dibromocycloalkane is colorless. To distinguish which bottle contains which compound, place a small quantity of each compound in a test tube and to each add a few drops of Br_2/CH_2Cl_2 solution. If the red color disappears, the compound is cyclohexene. If the red color remains, the compound is benzene.

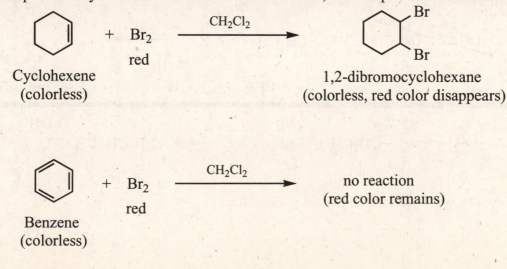

13.21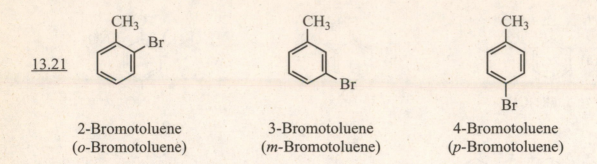

2-Bromotoluene   3-Bromotoluene   4-Bromotoluene
(*o*-Bromotoluene)   (*m*-Bromotoluene)   (*p*-Bromotoluene)

13.23 (a) Nitration using HNO_3/H_2SO_4 followed by sulfonation using H_2SO_4. The order of the steps may also be reversed.

(b) Bromination using $Br_2/FeCl_3$ followed by chlorination using $Cl_2/FeCl_3$. The order of the steps may also be reversed.

13.25 (a), (b), (f), (g), (h), and (i): True

(c) False: the pK_a of phenol (10) is larger than that of acetic acid (4.7).

(d) False: R-H is converted to R-OOH through autooxidation.

(e) False: carbon free radicals are uncharged.

13.27 Autooxidation is the reaction of a C-H group with oxygen, O_2, to form a hydroperoxide, R-OOH.

13.29

Step 2a

Step 2b

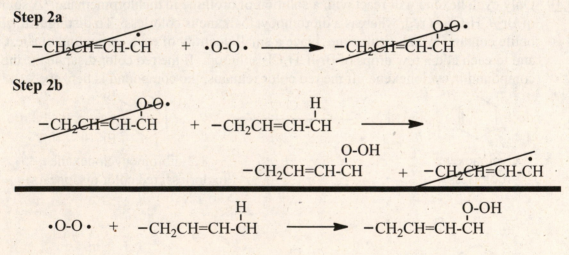

13.31 They all have a substituted phenol in common.

13.33 The functional groups for albuterol are circled and identified with the family of compounds they belong to.

(a)

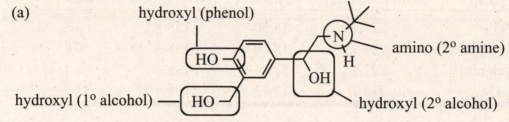

hydroxyl (phenol)

amino (2° amine)

hydroxyl (1° alcohol)

hydroxyl (2° alcohol)

(b) The phenolic proton is the most acidic in the molecule (pK$_a$ = 10).

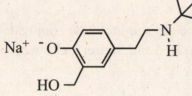

(c) The amino nitrogen is protonated by HCl to form a hydrochloride salt.

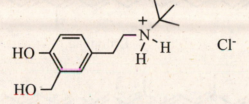

13.35 An advantage to using DDT as an insecticide includes its effectiveness in ridding large areas of the world of insect pests, which increased crop yields and prevented the spread of insect borne diseases such as malaria and typhus. The disadvantage of DDT was that it inhibited the mechanism by which birds incorporate calcium into their eggs. The eggs became too thin to survive incubation and the embryo died. This caused widespread declines in bird populations. DDT resists biodegradation and persists in the environment.

13.37 The following structure outlines the structure for dichlorodiphenyldichloroethylene.

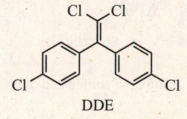

DDE

13.39 By definition, a carcinogen is a cancer-causing substance. The carcinogens present in cigarette smoke belong to the class of compounds called polynuclear aromatic hydrocarbons (PAHs).

13.41 Cyclonite (RDX) has the greatest mass percentage of nitrogen and NO_2 groups.

Explosive	MW (g/mol)	% mass N	NO_2 (g/mol)	% mass NO_2
TNT	227.1	18.50	138	60.8
Nitroglycerine	227.1	18.50	138	60.8
Cyclonite	222.1	37.84	138	62.1
PETN	316.1	17.72	184	58.2

13.43 The functional groups most responsible for the water solubility of these dyes are the two ionic $-SO_3^-Na^+$ groups in each.

13.45 (a) Lycopene and β-carotene are non-aromatic conjugated (alternating single and double bonds) polyenes. Allura Red and Sunset Yellow involve aromatic rings linked by a N=N bond.

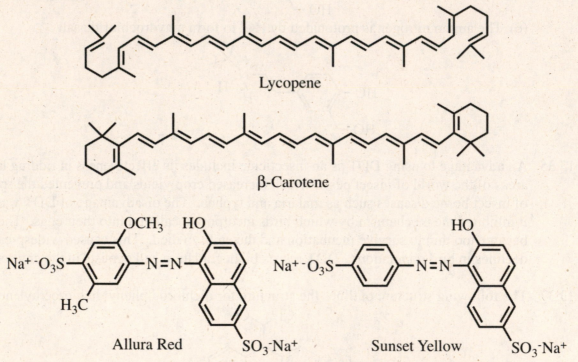

Lycopene

β-Carotene

Allura Red

Sunset Yellow

(b) All four compounds have multiple conjugated double bonds in common.

(c) Lycopene and β-carotene are hydrocarbons that are nonpolar and have no interaction with water. Sunset Yellow and Allura Red have two ionic $-SO_3^-$ groups attached to the aromatic rings that promote water solubility by forming strong interactions with water.

13.47

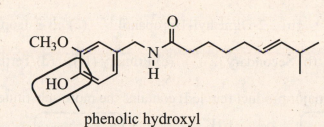

phenolic hydroxyl

13.49 Capsaicin is used to relieve pain associated with postherpetic neuralgia, a complication of shingles. It is also used to treat persistent foot and leg pain.

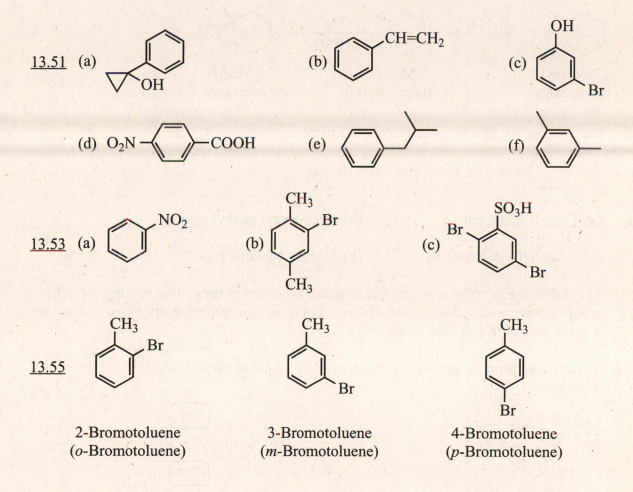

13.51 (a)

(b)   CH=CH₂

(c)   OH, Br

(d)   O₂N——COOH

(e)

(f)

13.53 (a)   NO₂

(b)   CH₃, Br, CH₃

(c)   SO₃H, Br, Br

13.55

CH₃, Br

2-Bromotoluene
(*o*-Bromotoluene)

CH₃, Br

3-Bromotoluene
(*m*-Bromotoluene)

CH₃, Br

4-Bromotoluene
(*p*-Bromotoluene)

14.1   (a) 2-Heptanol   (b) 2,2-Dimethyl-1-propanol   (c) *cis*-3-Isopropylcyclohexanol

14.2   (a) Primary (1°)   (b) Secondary (2°)   (c) Primary (1°)   (d) Tertiary (3°)

14.3   In each case, the major product (circled) contains the more substituted double bond.

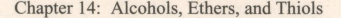

14.4

2-Methyl-   1-Methyl-   1-Methyl-
cyclohexanol   cyclohexene (C)   cyclohexanol (D)

14.5   Each secondary alcohol is oxidized to a ketone.

(a) [cyclohexanone structure] =O   (b) $CH_3\overset{O}{\overset{\|}{C}}CH_2CH_2CH_3$

14.6   (a) Ethyl isobutyl ether   (b) Cyclopentyl methyl ether

14.7   (a) 3-Methyl-1-butanethiol   (b) 3-Methyl-2-butanethiol

14.9   The difference is in the number of carbon atoms bonded to the carbon bearing the –OH group. For a primary alcohol, there is one. For a secondary alcohol, there are two and for a tertiary alcohol, there are three.

14.11   There are no primary alcohols in Problem 14.10. Both alcohols (a) and (b) are tertiary alcohols.

(a) [bicyclic structure with CH₃ and OH circled]

(b) $H_3C-\underset{\underset{CH_3}{|}}{\overset{\overset{CH_3}{|}}{C}}-OH$ (with CH₃ groups circled)

Three non-hydrogen groups (circled) attached to each carbon bonded to the hydroxyl. The alcohols in parts (c) and (d) are secondary alcohols.

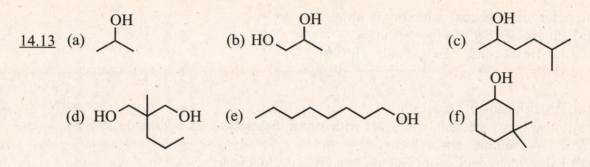

14.13 (a)

(b) HO

(c)

(d) HO

(e)

(f)

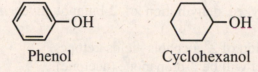

14.15 Phenol contains a benzene ring. The following are structural formulas for phenol and cyclohexanol.

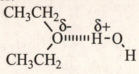

Phenol   Cyclohexanol

14.17 Low-molecular-weight alcohols form strong hydrogen bonds with water molecules, which cause them to dissolve in water. Low-molecular-weight alkanes are nonpolar compounds and do not dissolve in water because they are not capable of interacting with water molecules.

14.19 As the molecular weight of an alcohol increases, the nonpolar hydrocarbon component of the alcohol molecule becomes a larger fraction of the molecule's surface area. The physical properties of higher molecular weight alcohols, including solubility in water, then become more like those of the hydrocarbons of similar carbon skeletons.

14.21 The following illustration describes the hydrogen bonding between the oxygen of diethyl ether and the hydrogen of water.

$$CH_3CH_2$$
$$\backslash \delta- \quad \delta+$$
$$O\text{\tiny{IIIIII}}H-O$$
$$CH_3CH_2 \qquad \qquad H$$

14.23 In order of increasing boiling point, they are:
$$CH_3CH_2CH_2CH_3 \qquad CH_3CH_2OCH_2CH_3 \qquad CH_3CH_2CH_2OH$$
$$0°C \qquad\qquad\qquad 35°C \qquad\qquad\qquad 97°C$$

14.25 The thickness (viscosity) of these three liquids is related to the degree of hydrogen bonding between their molecules in the liquid state. Hydrogen bonding is strongest between molecules of glycerol, weaker between molecules of ethylene glycol, and weakest between molecules of ethanol.

14.27 In order of decreasing solubility in water, they are:
  (a) Ethanol > Diethyl ether > Butane
  (b) 1,2-Hexanediol > 1-Hexanol > Hexane

14.29 (a), (b), (g), and (i): True
  (c) False: only phenols will react with strong bases such as hydroxide to form water-soluble salts.
  (d) False: dehydration of cyclohexanol gives cyclohexene.
  (e) False: acid catalyzed dehydration of alcohols favors the alkene product with the most substituted double bond.
  (f) False: acid catalyzed dehydration of 2-butanol gives *trans*-2-butene as the major product
  (h) False: the oxidation of 2° alcohols yields ketones.
  (j) False: the product will be 2-propanone (acetone).

14.31 Phenols are weak acids, with pK_a values around 10. Alcohols are considerably weaker acids, having similar acidities as water, with pK_a values between 16 and 18.

14.33 The first reaction is an acid-catalyzed dehydration; the second is an oxidation.

(a) $$CH_3CH_2CH_2CH_2OH \xrightarrow[\text{heat}]{H_2SO_4} CH_3CH_2CH{=}CH_2 + H_2O$$

(b) $$CH_3CH_2CH_2CH_2OH \xrightarrow[H_2SO_4]{K_2Cr_2O_7} CH_3CH_2CH_2\overset{\displaystyle O}{\overset{\|}{C}}OH$$

14.35 Oxidation of a primary alcohol by $K_2Cr_2O_7/H_2SO_4$ gives a carboxylic acid.

(a) $$CH_3(CH_2)_6CH_2OH \xrightarrow[H_2SO_4]{K_2Cr_2O_7} CH_3(CH_2)_6\overset{\displaystyle O}{\overset{\|}{C}}OH$$

(b) $$HOCH_2CH_2CH_2CH_2OH \xrightarrow[H_2SO_4]{K_2Cr_2O_7} HO\overset{\displaystyle O}{\overset{\|}{C}}CH_2CH_2\overset{\displaystyle O}{\overset{\|}{C}}OH$$

14.37 Each can be prepared from 1-propanol (circled) as shown in this flow chart.

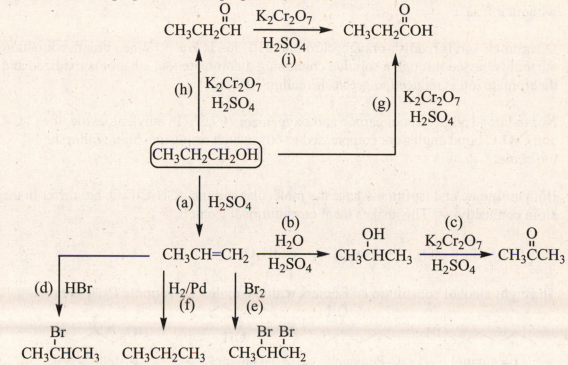

14.39 Ethanol and ethylene glycol are derived from ethylene. Ethanol is a solvent and is the starting material for the synthesis of diethyl ether, also an important solvent. Ethylene glycol is used in automotive antifreezes and is one of the two starting materials required for the synthesis of poly(ethylene terephthalate), better known as PET (Chapter 18).

14.41 (a): True
       (b) False: ethers are les soluble than their isomeric alcohols.
       (c) False: ethers are much less reactive than alcohols.

14.43 (a) Ethyl isopropyl ether   (b) Dibutyl ether   (b) Diphenyl ether

14.45 (a) 2-Butanethiol   (b) 1-Butanethiol   (c) Cyclohexanethiol

14.47 Because 1-butanol molecules associate in the liquid state by hydrogen bonding, it has the higher boiling point (117°C). There is very little polarity to an S-H bond. The only interactions among 1-butanethiol molecules in the liquid state are the considerably weaker London dispersion forces. For this reason, 1-butanethiol has the lower boiling point (98°C).

14.49 (a), (b), (c), (d), (e), and (f): True
       no false statements

14.51 Nobel discovered that diatomaceous earth absorbs nitroglycerin so that it will not explode without a fuse.

14.53 Dichromate ion is reddish-orange; chromium(III) ion is green. When breath containing ethanol is passed through a solution containing dichromate ion, ethanol is oxidized and dichromate ion is reduced to green chromium(III) ion.

14.55 Normal bond angles about carbon and oxygen are 109.5°. In ethylene oxide, the C-C-O and C-O-C bond angles are compressed to 60°, which results in strain within the molecule.

14.57 Both enflurane and isoflurane have the molecular formula $C_3H_2ClF_5O$, but differ in their atom connectivity. This makes them constitutional isomers.

14.59 $CH_3CH_2OH + 3O_2 \longrightarrow 2CO_2 + 3H_2O$

14.61 The eight alcohol constitutional isomers with the molecular formula $C_5H_{12}O$ are:

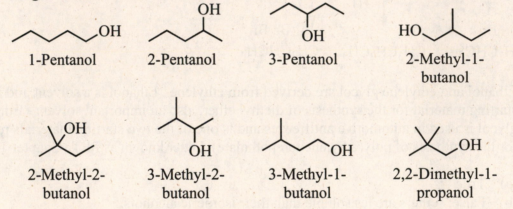

1-Pentanol	2-Pentanol	3-Pentanol	2-Methyl-1-butanol
2-Methyl-2-butanol	3-Methyl-2-butanol	3-Methyl-1-butanol	2,2-Dimethyl-1-propanol

14.63 Ethylene glycol has two -OH groups by which each molecule participates in hydrogen bonding, whereas 1-propanol has only one. The stronger intermolecular forces of attraction between molecules of ethylene glycol give it the higher boiling point.

14.65 Arranged in order of increasing boiling point, they are:

$CH_3CH_2CH_2CH_2CH_2CH_3$ $CH_3CH_2CH_2CH_2CH_2OH$ $HOCH_2CH_2CH_2CH_2OH$

Hexane	1-Pentanol	1,4-Butanediol
(bp 69°C)	(bp 138°C)	(bp 230°C)

14.67 The solvent with the greater solubility in water is circled.

(a) CH_2Cl_2 or $\boxed{CH_3CH_2OH}$ (b) $CH_3CH_2OCH_2CH_3$ or $\boxed{CH_3CH_2OH}$

14.69 Each can be prepared from 2-methylcyclohexanol (circled) as shown in this flow chart.

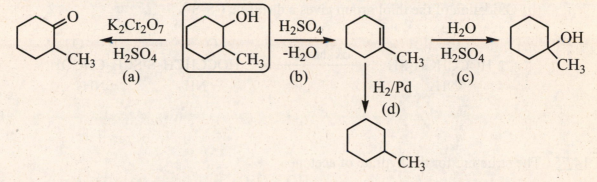

14.71 In part (c), assume that the two possible alcohols formed can be separated.

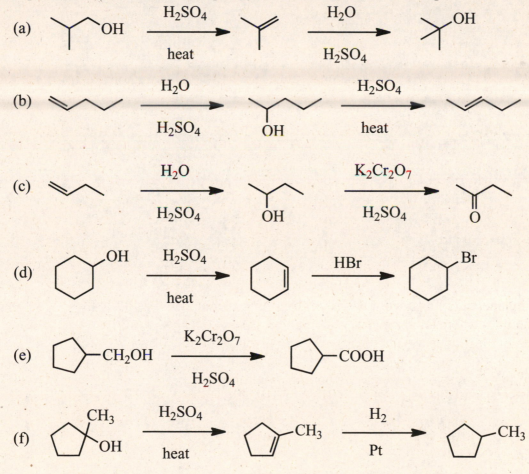

14.73 (a) The three functional groups are a thiol, a primary amine, and a carboxylic acid.
(b) Oxidation of the thiol group gives a disulfide (-S-S-).

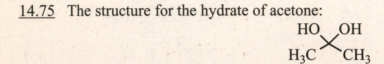

14.75 The structure for the hydrate of acetone:

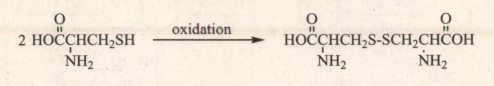

Chapter 15: Chirality The Handedness of Molecules

15.1 The enantiomers of each part are drawn with two groups in the plane of the paper, a third group toward you in front of the plane, and the fourth group away from you behind the plane.

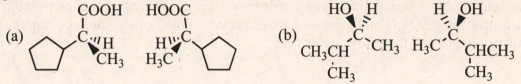

15.2 The group of higher priority in each set is circled.

(a) $\boxed{-CH_2OH}$ and $-CH_2CH_2COOH$ (b) $\boxed{-CH_2NH_2}$ and $-CH_2COOH$

15.3 The order of priorities and the configuration is shown in the drawings.

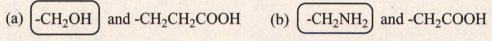

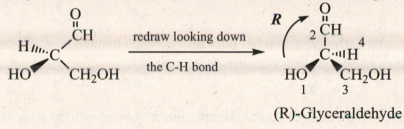

(R)-Glyceraldehyde

15.4 (a) Compounds 1 and 3 are one pair of enantiomers. Compounds 2 and 4 are a second pair of enantiomers.
(b) Compounds 1 and 2, 1 and 4, 2 and 3, and 3 and 4 are diastereomers.

15.5 Four stereoisomers are possible for 3-methylcyclohexanol. The *cis* isomers are one pair of enantiomers and the *trans* isomers are a second pair of enantiomers.

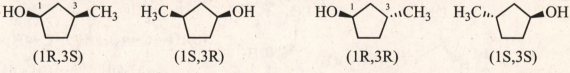

     (1R,3S)   (1S,3R)   (1R,3R)   (1S,3S)

15.6 Stereocenters are identified with an asterisk and the number of stereoisomers possible is shown under the structural formula.

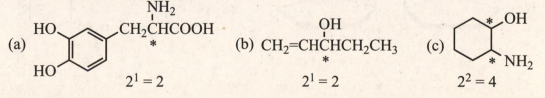

129

15.7   (a), (b), (c), (e), (f), (g), (h), and (i): True
(d) False: constitutional isomers differ in their atom connectivity.

15.9   An achiral object is one that has no handedness; it is an object that is superposable upon its mirror image. H_2O and 1-butanol are examples of molecules that have superposable mirror images and therefore, are achiral.

15.11   Both constitutional isomers and stereoisomers have the same molecular formula. Whereas stereoisomers have the same connectivity, constitutional isomers have a different connectivity.

15.13   2-Pentanol has a stereocenter (carbon 2). 3-Pentanol has no stereocenter.

15.15   The carbon of a carbonyl group has only three groups bonded to it. To be a stereocenter, a carbon must have four different groups bonded to it.

15.17   Compounds (b), (c), and (d) contain stereocenters, here identified with asterisks.

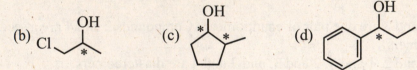

15.19   Following are mirror images of each molecule.

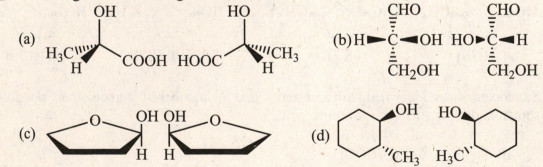

15.21 (a), (c), (e), and (f): True
 (b) False: the equation is 2^n = stereoisomers possible, where n = the number of stereocenters. Therefore, a molecule with three stereocenters has $2^3 = 8$ stereoisomers possible.
 (d) False: 3-pentanol is achiral.

15.23 Parts (b) and (c) contain stereocenters.

15.25 Each stereocenter is identified with an asterisk. Under each structural formula is the maximum number of stereoisomers possible. Compounds (a), (c), and (d) have stereocenters that are chiral carbons. Compound (b) has a double bond that is a stereocenter (cis and trans stereoisomers possible).

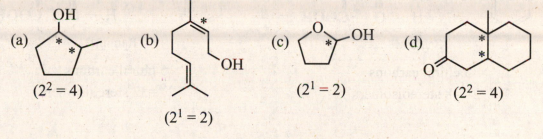

15.27 The optical rotation of its enantiomer is +41°.

15.29 (a), (b), and (d): True
 (c) False: cis/trans stereoisomers do not need to be chiral, therefore there is no requirement that they need to be optically active.

15.31 All three structures are chiral. The chiral carbons are identified with an asterisk. The atoms and bonds in bold are those structural feature similarities with Captopril.

Captopril

2 chiral carbons

$2^2 = 4$ stereoisomers

Quinapril

3 chiral carbons

$2^3 = 8$ stereoisomers

Enalopril

3 chiral carbons

$2^3 = 8$ stereoisomers

Ramipril

5 chiral carbons

$2^5 = 32$ stereoisomers

All of the molecules share the structural features similar to Captopril:

Quinapril, Enalopril, and Ramipril share this structural feature:

Enalopril and Ramipril share this structural feature:

Enalopril and Ramipril share this structural feature with Captopril:

15.33 The two stereocenters are identified with asterisks.

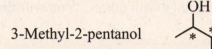

3-Methyl-2-pentanol

15.35 Each stereocenter is identified with an asterisk. Under the name of each compound is the number of stereoisomers possible for it.

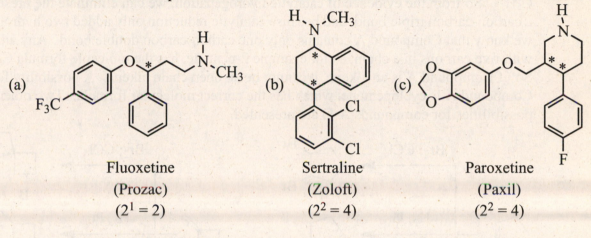

(a) Fluoxetine (Prozac) $(2^1 = 2)$

(b) Sertraline (Zoloft) $(2^2 = 4)$

(c) Paroxetine (Paxil) $(2^2 = 4)$

15.37 (a) The only possible compound that does not show *cis-trans* isomerism and has no stereocenter is 1-methylcyclohexanol.

1-Methylcyclohexanol

(b) Only 4-methylcyclohexanol shows *cis-trans* isomerism but has no stereocenter. Drawn here are the *cis* and *trans* isomers.

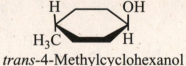

trans-4-Methylcyclohexanol   *cis*-4-Methylcyclohexanol

(c) Both 2-methylcyclohexanol and 3-methylcyclohexanol have two stereocenters and can exist as *cis* and *trans* isomers. The stereocenters in each are identified with asterisks.

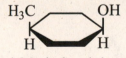

2-Methylcyclohexanol   3-Methylcyclohexanol

15.39 The majority have a right-handed twist because the machines that make them all impart the same twist. Share your findings with others. You will find it interesting to compare them.

15.41 Compound **A** could contain either a carbon triple bond, but if it did, treatment with H_2 in the presence of a transition metal catalyst would add two moles of hydrogen to give C_5H_{12}. So from the evidence of catalytic hydrogenation, we can eliminate the presence of a carbon-carbon triple bond, and because catalytic reduction only added two hydrogens, we know that Compound **A** contains only one carbon-carbon double bond. Any alkene with five carbons in a chain, as for example 2-pentene, has the molecule formula C_5H_{10}. So, if Compound **A** is an alkene, it cannot be an open chain alkene. A possibility for Compound **A** is cyclopentene, which has the correct molecular formula. Two other possibilities for compounds A-D are presented.

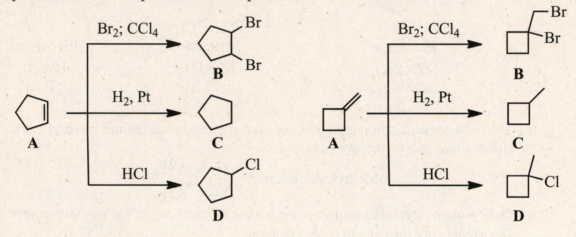

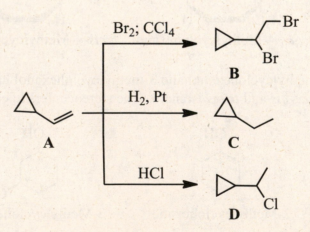

15.43 (a) The eight stereocenters are identified with asterisks.

(b) There are $2^8 = 256$ stereoisomers and 128 pairs of enantiomers possible.

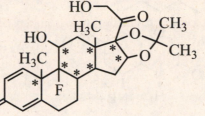

Triamcinolone acetonide

15.45 (a) See following structure:

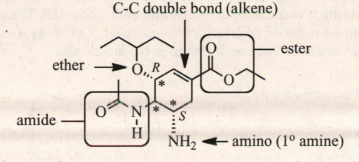

Oseltamivir (Tamaflu)

(b) Its molecular formula is $C_{16}H_{28}N_2O_4$.

(c) It is chiral, and if it were isolated from a natural source, it would be optically active and would rotate the plane of polarized light. If, however, it were synthesized in the laboratory, in the absence of chiral reagents and catalysts, it would be a mixture of stereoisomers, racemic, and optically inactive.

(d) Three chiral centers, only two defined by the given structure.

(e) *Oseltamivir* and its enantiomer:

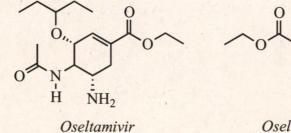
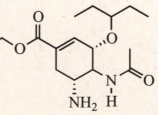

Oseltamivir   *Oseltamivir* enantiomer

(f) Drawing a diastereomers involves inverting only one stereocenter while leaving the others the same. Examples include:

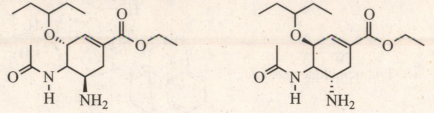

(g) The stereochemistry of the carboxyl group is undefined in the problem, so lets consider the two possibilities. The orientation of each substituent on the six-membered ring is marked (e) for equatorial or (a) for axial. The most stable conformation for the two diastereomers is circled. Generally, the six-membered ring conformer with more equatorial groups is the most stable.

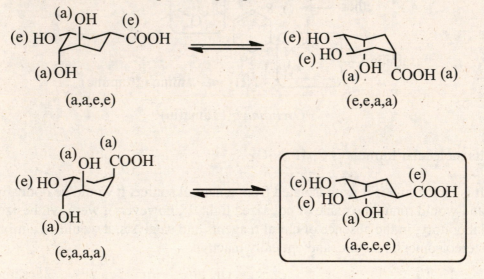

(h) *Oseltamivir* has three chiral centers giving it $2^3 = 8$ possible stereoisomers.

Chapter 16: Amines

16.1   Pyrrolidine has nine hydrogen atoms; its molecular formula is C_4H_9N. Purine has four hydrogen atoms; its molecular formula is $C_5H_4N_4$.

16.2   Following is a line-angle formula for each compound.

(a) 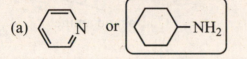   (b)   (c)

16.3   Following is a line-angle formula for each compound.

(a)   (b)   (c)

16.4   The stronger base is circled.

(a)   Aliphatic amines are stronger bases than heterocyclic aromatic amines.

(b)   Aliphatic amines are stronger bases than aromatic amines.

16.5   The product of each reaction is an amine salt.

(a) $(CH_3CH_2)_3\overset{+}{N}H$ Cl^-
Triethylammonium chloride

(b)   Piperidinium acetate

16.7   The classification of primary, secondary, and tertiary amines and alcohols is based on point of reference. In the case of alcohols, the terms 1°, 2°, and 3° refer to the number of carbons bonded to the carbon bearing the –OH group. In the case of amines, the terms 1°, 2°, and 3° refer to the number of carbons bonded to the nitrogen atom of the amine.

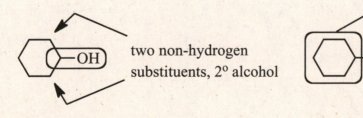

one non-hydrogen substituent on the amine nitrogen, 1° amine

two non-hydrogen substituents, 2° alcohol

16.9   Each compound has a six-membered ring with three double bonds.

16.11 Following is a structural formula for each amine:

(a) [structure: sec-butylamine with NH₂]

(b) $CH_3(CH_2)_6CH_2NH_2$

(c) [structure: neopentyl amine with NH₂]

(d) $H_2N(CH_2)_5NH_2$

(e) [structure: benzene ring with NH₂ and Br]

(f) $(CH_3CH_2CH_2CH_2)_3N$

16.13 Each amine is classified by type:

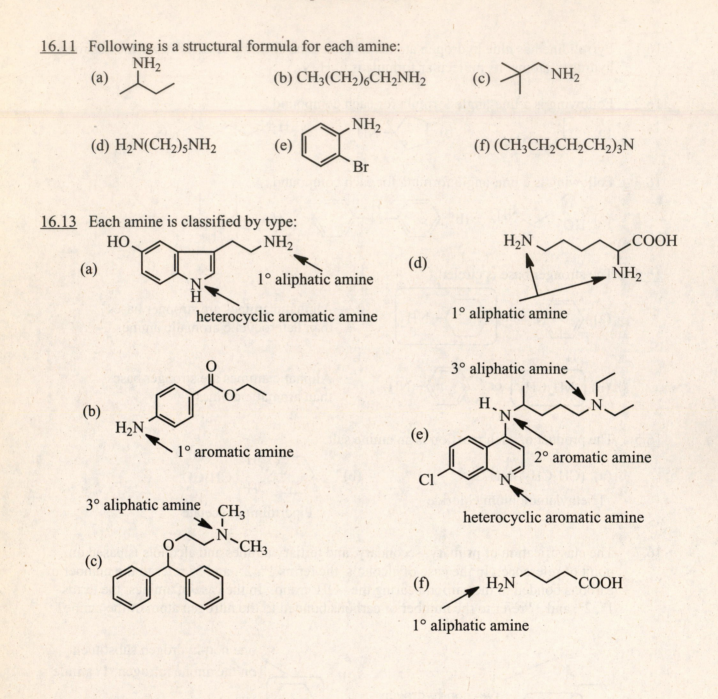

(a) 1° aliphatic amine
heterocyclic aromatic amine

(b) 1° aromatic amine

(c) 3° aliphatic amine

(d) 1° aliphatic amine

(e) 3° aliphatic amine
2° aromatic amine
heterocyclic aromatic amine

(f) 1° aliphatic amine

16.15 Following are structural formulas and names for the eight primary amines with molecular formula $C_5H_{13}N$. Of these, three are chiral; their stereocenters are identified with asterisks.

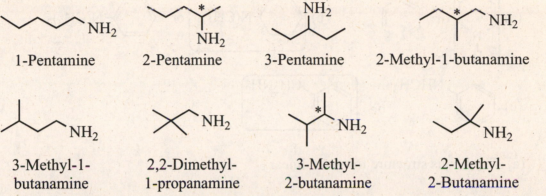

| 1-Pentamine | 2-Pentamine | 3-Pentamine | 2-Methyl-1-butanamine |

| 3-Methyl-1-butanamine | 2,2-Dimethyl-1-propanamine | 3-Methyl-2-butanamine | 2-Methyl-2-Butanamine |

16.17 Both propylamine (a 1° amine) and ethylmethylamine (a 2° amine) have an N-H group and hydrogen bonding occurs between their molecules in the liquid state. Because of this intermolecular force of attraction, these two amines have higher boiling points than trimethylamine, which has no N-H bond and, therefore, cannot participate in intermolecular hydrogen bonding.

16.19 2-Methylpropane is a nonpolar hydrocarbon and the only attractive forces between its molecules in the liquid state are the very weak London dispersion forces. Both 2-propanol and 2-propanamine are polar molecules and associate in the liquid state by hydrogen bonding. Hydrogen bonding is stronger between alcohol molecules than between 1° and 2° amine molecules because of the greater strength of an O-H----O hydrogen bond compared to an N-H----N hydrogen bond. It takes more energy to separate alcohol molecules in the liquid state from their neighbors than to separate amine molecules from their neighbors and, therefore, the alcohol has the higher boiling point.

16.21 (a), (b), (d), (e), (f), and (g): True
(c) False: amines are weaker bases than NaOH and KOH.

16.23 Structural formula for each amine salt:

(a) $(CH_3)_3\overset{+}{N}CH_2CH_3\ OH^-$

(b) $(CH_3)_2NH_2^+\ I^-$

(c) $(CH_3)_4N^+\ Cl^-$

(d) ⬡—$NH_3^+\ Br^-$

16.25 The stronger base is circled. The determining factor is that aliphatic amines are stronger bases than aromatic amines and heterocyclic aromatic amines.

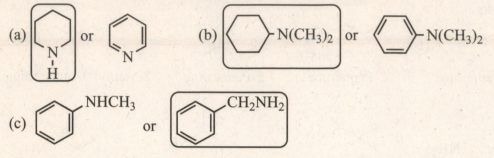

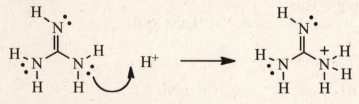

16.27 (a) The Lewis structure for guanidine:

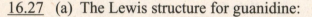

(b) Three equivalent contributing structures:

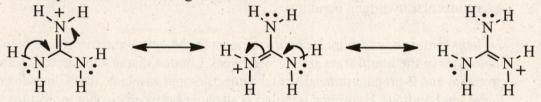

(c) Protonation of the –NH$_2$ nitrogen leads to a conjugate acid that is less stable because it lacks resonance stabilization. Protonation of the C=NH nitrogen results in a resonance stabilized conjugate acid as in (b). Acid-base equilibria favor the side with the weaker (more stable) acids and bases, therefore, protonation of the C=NH is favored.

no resonance stabilization

(d) The central carbon atom has three bonding pairs of electrons and no non-bonded electrons surrounding it, therefore, the VSEPR theory predicts a N-C-N bond angle of 120°.

(e) As the value for pKa decreases, the acid strength increases. Ammonium (NH$_4^+$, pK$_a$ = 9.3) is a stronger acid than the guanidinium ion (pKa = 13.6).

16.29 Treat each with dilute aqueous HCl. The amine will react with HCl to form a water-soluble salt. The alcohol does not react with this solution and is insoluble in it. Thus, the amine dissolves in aqueous HCl; the alcohol does not.

16.31 The primary aliphatic amine is the stronger base and forms the salt with HCl. The salt is named pyridoxamine hydrochloride.

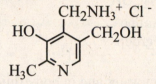

Pyridoxamine hydrochloride

16.33 (a) Epinephrine has two phenolic –OH groups, whereas amphetamine has none. Epinephrine has a 2° alcohol on its carbon side chain whereas amphetamine has none. Epinephrine is a 2° amine, whereas amphetamine is a 1° amine. Finally, both epinephrine and amphetamine are chiral, but their stereocenters are in different locations within the molecule.
(b) Methamphetamine is a 2° aliphatic amine whereas amphetamine is a 1° aliphatic amine. Both compounds are chiral at the same carbon.

16.35 Alkaloids are basic nitrogen-containing compounds found in the roots, bark, leaves, berries, or fruits of plants. In almost all alkaloids, the nitrogen atom is present as a member of a ring; that is, it is present in a heterocyclic ring. By definition, alkaloids are nitrogen-containing bases and, therefore, turn red litmus blue.

16.37 The tertiary aliphatic amine in the five-membered ring is the stronger base; therefore it will be protonated by the HCl.

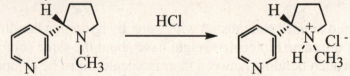

16.39 The common structural feature of benzodiazepines is a benzene ring fused to a seven-membered ring containing two nitrogen atoms.

<u>16.41</u> The structural formula of 4-aminobutanoic acid is drawn on the left showing the 1°
amino group (a base) and the carboxyl group (an acid). It is drawn on the right as an
internal salt.

a primary amine

a carboxylic acid

$H_2NCH_2CH_2CH_2\overset{\overset{O}{\parallel}}{C}OH$

un-ionized amino and carboxyl groups

$\overset{+}{H_3}NCH_2CH_2CH_2\overset{\overset{O}{\parallel}}{C}O^-$

internal salt

<u>16.43</u> Their salts are more soluble in water and in body fluids, and are more stable (less
reactive) toward oxidation by atmospheric oxygen.

<u>16.45</u> · Following is structural formula for each part.

(a) ⬡—NHCH₃ (b) ⬡—$\overset{\overset{CH_3}{|}}{N}CH_3$ (c) ⬡—CH₂NH₂ (d) $CH_3\overset{\overset{NH_2}{|}}{C}HCH_2CH_3$

(e) [pyrrolidine ring with N-CH₃]

(f) H_3C—⬡(with CH₃ groups)—NH₂

(g) $CH_3\overset{\overset{Cl^-\ CH_2CH_3}{\overset{+|}{}}}{N}CH_2CH_2CH_3$ with CH₃CHCH₃

<u>16.47</u> (a) CH₃SH is the strongest acid.
(b) (CH₃)₂NH is the strongest base.
(c) CH₃OH has the highest boiling point because it has the strongest hydrogen bonding.
(d) Molecules of CH₃OH form the strongest hydrogen bonds because the O-H bond is
more polar than the N-H and S-H bonds.

<u>16.49</u> Both alcohols and amines can interact with water by hydrogen bonding. Because amines
and alcohols of the same molecular weight have about the same solubility in water, the
strength of hydrogen bonding between their molecules must be comparable.

<u>16.51</u> (a) The following is the structure of 1-phenyl-2-amino-1-propanol and the reaction with
HCl to form its hydrochloride salt.

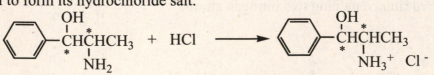

(b) 1-Phenyl-2-amino-1-propanol has two stereocenters, identified with asterisks, and
$2^2 = 4$ stereoisomers are possible.

16.53 The zwitter ion (A) is a better representation of gabapentin's structure. Equilibria favor the side with the weaker acid/weaker base.

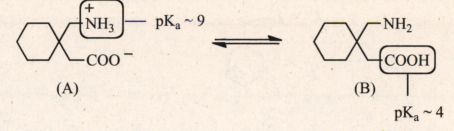

(A)   (B)

16.55 (a) The 2° aliphatic amine is more basic than the heterocyclic aromatic amine.
(b) The three stereocenters are identified with asterisks.

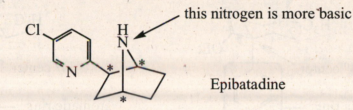

this nitrogen is more basic

Epibatadine

16.57 Alanine can be represented as the amino acid structure or the internal salt structure, which is better description of amino acids.

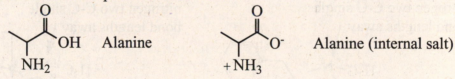

Alanine   Alanine (internal salt)

16.59 (a) The following skeleton is the structural feature common to Meperidine, Methadone, and Propoxyphene:

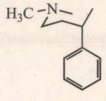

(b) The features that are consistent with the Beckett-Casey rules are outlined.

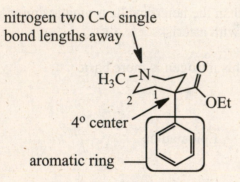

Meperidine

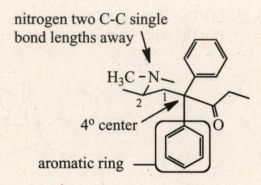

Methadone

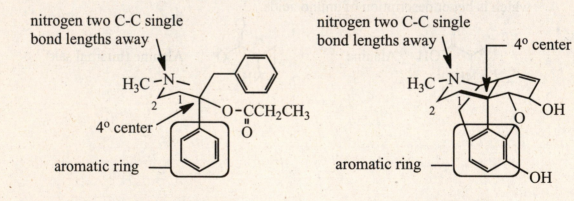

Propoxyphene

Morphine

Chapter 17: Aldehydes and Ketones

17.1 (a) 3,3-Dimethylbutanal (b) Cyclopentanone (c) 1-Phenyl-1-propanone

17.2 Following are line-angle formulas for each aldehyde with the molecular formula $C_6H_{12}O$. In the three aldehydes that are chiral, the stereocenter is identified with an asterisk.

Hexanal   4-Methylpentanal   3-Methylpentanal   2-Methypentanal

2,3-Dimethylbutanal   3,3-Dimethylbutanal   2,2-Dimethylbutanal   3-Ethybutanal

17.3 (a) 2,3-Dihydroxypropanal (b) 2-Aminobenzaldehyde (c) 5-Amino-2-pentanone

17.4 Each aldehyde is oxidized to a carboxylic acid.

Hexanedioic acid
(Adipic acid)

3-Phenylpropanoic acid

17.5 Each primary alcohol comes from reduction of an aldehyde. Each secondary alcohol comes from reduction of a ketone.

(b) CH_3O—⟨⟩—$CH_2\overset{\text{O}}{\overset{\|}{C}}H$

17.6 Shown first is the hemiacetal and then the acetal.

Benzaldehyde   Hemiacetal   Acetal

17.7 (a) A hemiacetal formed from 3-pentanone (a ketone) and ethanol.
(b) Neither a hemiacetal nor an acetal. This compound is the dimethyl ether of ethylene glycol.
(c) An acetal derived from 5-hydroxypentanal and methanol.

17.8   Following is the keto form of each enol.

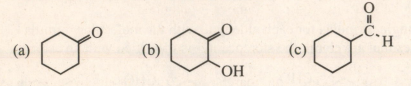

(a)   (b)   (c)

17.9   (a), (b), and (c): True
(d)   False: carbonyl carbons can only have three substituents. Therefore it cannot be a
        stereocenter.

17.11   In an aromatic aldehyde, the –CHO group is bonded to a benzene ring. In an aliphatic
        aldehyde, it is bonded to tetrahedral carbon atom.

17.13   Compounds (b), (c), (d), and (f) each contain a carbonyl group.

17.15   Of the four aldehydes with molecular formula $C_5H_{10}O$, only one is chiral. Its stereocenter
        is identified with an asterisk.

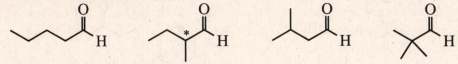

17.17   The following are structural formulas for each aldehyde.

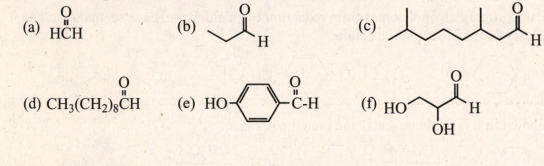

(a) HCH   (b)   (c)

(d) $CH_3(CH_2)_8CH$ (e) HO——C-H   (f) HO   H
                                                        OH

17.19   (a) 4-heptanone   (b) 2-Methylcyclopentanone
        (c) cis-2-Methyl-2-butenal   (d) 2-Hydroxypropanal
        (e) 1-Phenyl-2-propanone   (f) Hexanedial

17.21 (a) Numbering the parent chain from the end giving the carbonyl the lowest possible number reveals the structure's correct name is 2-butanone.

O
1   3   2-butanone   not   4   2
  2   4   3   1

(b) When a carbonyl is on the end of the parent chain, it is an aldehyde, not a ketone. Ketones have an –*one* ending and aldehydes have an –*al* ending. The correct name is butanal.

O
4   2   H   butanal
  3   1

(c) The longest parent chain is five carbons, not four, therefore, the correct name is pentanal, not 4-methylbutanal.

O
4   2   H   pentanal
  3   1
5

(d) Numbering the parent chain from the end giving the carbonyl the lowest possible number reveals that the parent chain is 2-butanone and the structure's correct name is 3,3-dimethyl-2-butanone.

O   O
1   3   3,3-dimethyl-2-butanone   not   4   2
  2   4   3   1

17.23 (a), (b), (c), and (d): True
no false statements

17.25 Because it is a hydrogen bond acceptor, the carbonyl group is hydrophilic and it promotes water solubility. The hydrocarbon portions of each molecule are hydrophobic; they decrease solubility in water. For acetone, the hydrophilic character of the carbonyl group outweighs the hydrophobic character of its two methyl groups, and acetone is soluble in water. For 4-heptanone, the hydrophobic character of the two propyl groups outweighs the hydrophilic character of its carbonyl group, and 4-heptanone is insoluble in water.

17.27 Pentane is a nonpolar hydrocarbon and the only attractive forces between its molecules in the liquid state are the very weak London dispersion forces. Pentane, therefore, has the lowest boiling point. Pentanal and 1-butanol are both polar molecules. Because 1-butanol has a polar -OH group and its molecules can associate by hydrogen bonding, the intermolecular attractions between molecules of 1-butanol are greater than those between molecules of pentanal. 1-Butanol, therefore, has a higher boiling point than pentanal.

17.29 Acetone has no –OH or –NH groups through which to form intermolecular hydrogen bonds between its molecules.

17.31 Aldehydes are oxidized by potassium dichromate in sulfuric acid to carboxylic acids. Ketones are not oxidized under these conditions. Secondary alcohols are oxidized to ketones.

(a) $CH_3CH_2CH_2\overset{\overset{\displaystyle O}{\|}}{C}OH$ (b) [benzene ring]$-\overset{\overset{\displaystyle O}{\|}}{C}OH$ (c) no reaction   (d) [cyclohexanone]$=O$

17.33 (a) Treat each with Tollens' reagent. Only pentanal will give a silver mirror.
(b) Treat each with $K_2Cr_2O_7/H_2SO_4$. Only 2-pentanol is oxidized (to 2-pentanone), which causes the red color of $Cr_2O_7^{2-}$ ion to disappear and be replaced by the green color of Cr^{3+} ion.

17.35 The white solid is benzoic acid, formed by air oxidation of benzaldehyde.

17.37 These experimental conditions reduce an aldehyde to a primary alcohol and a ketone to a secondary alcohol. Both (a) and (c) are chiral. Each stereocenter is identified with an asterisk.

(a) $CH_3\overset{\overset{\displaystyle OH}{|}}{\underset{\displaystyle *}{C}}HCH_2CH_3$ (b) $CH_3(CH_2)_4CH_2OH$ (c) [cyclopentane ring with OH* and *CH$_3$]   (d) [benzene ring with CH_2OH and OH]

17.39 (a) The following is the structural formula for 1,3-dihydroxy-2-propanone.

$$HOCH_2\overset{\overset{\displaystyle O}{\|}}{C}CH_2OH$$

1,3-Dihydroxy-2-propanone
(Dihydroxyacetone)

(b) Because it has two hydroxyl groups and one carbonyl group, all of which can interact with water molecules by hydrogen bonding, predict that it is soluble in water.
(c) Its reduction gives 1,2,3-propanetriol, better known as glycerol or glycerin.

$$HOCH_2\overset{\overset{\displaystyle O}{\|}}{C}CH_2OH \xrightarrow[\text{H}_2\text{O}]{\text{NaBH}_4} HOCH_2\overset{\overset{\displaystyle OH}{|}}{C}HCH_2OH$$

1,2,3-Propanetriol

17.41 The first two reactions are reduction. There is no reaction in (c) or (d); ketones are not oxidized by these reagents.

(a,b) [benzene ring]$-\overset{\overset{\displaystyle OH}{|}}{C}HCH_3$

17.43 For an aldehyde or ketone to undergo keto-enol tautomerism, there must be a C-H bond adjacent to the carbonyl. Compounds (a), (b), (d), and (f) satisfy this requirement and will undergo keto-enol tautomerism.

17.45 Following are the keto forms of each enol.

(a) [cyclopentanone structure with =O]

(b) $CH_3\overset{O}{\overset{\|}{C}}CH_2CH_2CH_2CH_3$

(c) [benzene ring]$-CH_2\overset{O}{\overset{\|}{C}}CH_3$

17.47 A hemiacetal contains a carbon atom bonded to one –OH group and one –OR group, where R may be an alkyl or aryl group. An acetal contains a carbon atom bonded to two –OR groups, where R may be alkyl or aryl.

17.49 (a) hemiacetal   (b) acetal   (c) acetal
      (d) hemiacetal   (e) cyclic acetal   (f) acetal

17.51 Following are structural formula for the product of each hydrolysis.

(a) $CH_3CH_2\overset{O}{\overset{\|}{C}}CH_2CH_3$ + HO⌒⌒OH

(b) [benzene ring with]CHO and OCH$_3$ + 2CH$_3$OH

(c) [cyclohexane ring]=O + 2CH$_3$OH

(d) [cyclohexane ring with] –OH and C=O with H + CH$_3$OH

17.53 (a) The two C=C double bonds capable of *cis/trans* isomerism are identified with an asterisk. Both C=C double bonds have *trans* stereochemistry.

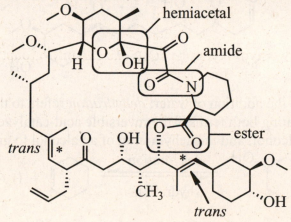

(b) Each stereocenter is identified with an asterisk. There are 16 stereocenters (14 chiral centers and two double bonds) with $2^{16} = 65536$ possible stereoisomers.

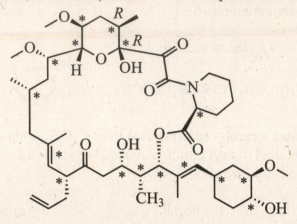

(c) There are no aromatic components in FK-506.
(d) Both of the chiral centers have the R configuration. See solution 17.53b for locations.
(e) The hemiacetal and ester are circled in solution 17.53a.
(f) The hemiacetal hydrolysis product retains a macrocycle structure:

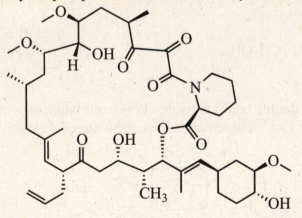

17.55 *Hydration* refers to the addition of water; *dehydration* refers to the elimination of water. An example illustrating both terms is the reversible acid-catalyzed addition of H_2O to an alkene to give an alcohol, and the dehydration of an alcohol to give an alkene.

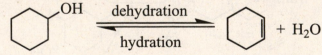

17.57 Compounds (a), (b), and (c) can be formed by reduction of the aldehyde or ketone shown. Compound (d) is a 3° alcohol and cannot be formed in this manner.

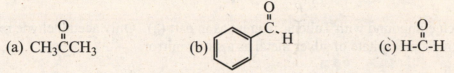

(a) $CH_3\overset{\overset{\displaystyle O}{\|}}{C}CH_3$ (b)   (c) $H-\overset{\overset{\displaystyle O}{\|}}{C}-H$

17.59 The alkene of three carbons is propene. According to Markovnikov's rule, H^+ adds to the carbon of the double bond that already has the greater number or hydrogens, which in the cases of propene is carbon-1 to give a 2° carbocation to which a molecule of water adds. Acid-catalyzed hydration of this alkene gives 2-propanol as the major product, not 1-propanol.

17.61 Each conversion can be brought about by acid-catalyzed hydration of the alkene to a secondary alcohol, followed by oxidation of the secondary alcohol to a ketone.

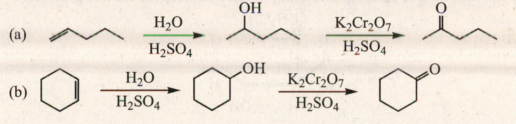

(a)

(b)

17.63 The carbonyl functional group for each aldehyde and ketone is circled.

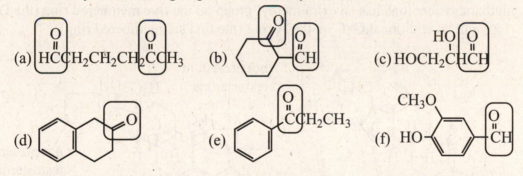

(a) $HC\!CH_2CH_2CH_2CCH_3$ (b)   (c) $HOCH_2CHCH$

(d)   (e)   (f) HO

17.65 Formulas for the one ketone and two aldehydes with molecular formula C_4H_8O are:

(a) $CH_3\overset{\overset{\displaystyle O}{\|}}{C}CH_2CH_3$ (b) $CH_3CH_2CH_2\overset{\overset{\displaystyle O}{\|}}{C}H$ and $CH_3\overset{\overset{\displaystyle O}{\|}}{C}HCH$
$\qquad\qquad\qquad\qquad\qquad\qquad\qquad\qquad\qquad\qquad\qquad\quad |$
$\qquad\qquad\qquad\qquad\qquad\qquad\qquad\qquad\qquad\qquad\quad CH_3$

17.67 2-Propanol has the higher boiling point because of the greater attraction between its molecules due to hydrogen bonding through hydroxyl groups.

17.69 (a) Treat each compound with Tollens' reagent. Only benzaldehyde reduces Ag^+ to give a precipitate of silver metal as a silver mirror.

(b) Treat each compound with Tollens' reagent as in part (a). Only acetaldehyde reduces Ag^+ to give a precipitate of silver metal as a silver mirror.

17.71 Shown in the equation are structural formulas for the equilibrium products.

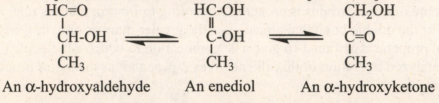

An α-hydroxyaldehyde   An enediol   An α-hydroxyketone

17.73 (a) Carbon 5 provides the –OH group, and carbon 1 provides the –CHO group.
(b) The following is a structural formula for the free aldehyde form of glucose.

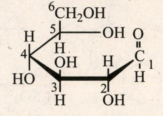

17.75 (a) The difference between testosterone and methandrostenolone is that methandrostenolone has an extra methyl group on the five membered ring (the D ring) and an additional C=C on the A ring (the first six membered ring).

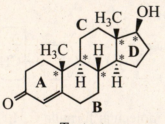

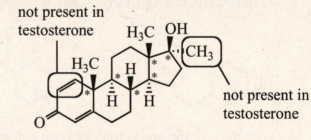

Testosterone   Methandrostenolone
$2^6 = 64$ possible stereoisomers $2^6 = 64$ possible stereoisomers

(b) Both testosterone and methandrostenolone have six stereocenters with the possibility of $2^6 = 64$ stereoisomers each.

17.77 Secondary alcohols oxidize to ketones. The oxidation of 1° alcohols such as in part (b) are difficult to stop after the first step and usually proceed onto carboxylic acids.

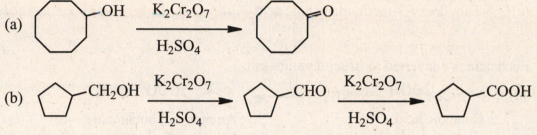

18.1   Glyceric acid and mevalonic acid are chiral.
        (a) 2,3-Dihydroxypropanoic acid   (b) 3-Aminopropanoic acid
        (c) 3,5-Dihydroxy-3-methylpentanoic acid

18.2   Each acid is converted to its ammonium salt.

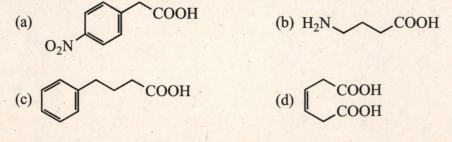

    (a)   COOH $+ NH_3$ ⟶ $COO^-\ NH_4^+$

              Butanoic acid   Ammonium butanoate
             (Butyric acid)   (Ammonium butyrate)

    (b)   OH $+ NH_3$ ⟶   OH

                 COOH $COO^-\ NH_4^+$

      (S)-2-Hydroxypropanoic   Ammonium (S)-2-hydroxypropanoate
              acid   (Ammonium (S)-lactate)
            (S)-Lactic acid)

18.3   (a) [structure] $+ H_2O$ (b) [structure] $+ H_2O$

18.5   (a), (b), (c), (e), (f), (g), and (h): True
        (d) False: Carbon #3 has two of the same groups (methyls) and there are no other
        stereocenters. 3-Methylbutanoic acid is achiral.

18.7   (a) 3,4-Dimethylpentanoic acid   (b) 2-Aminobutanoic acid   (c) Hexanoic acid

18.9   Following are structural formulas for each carboxylic acid.
        (a) [structure] COOH, O_2N (b) H_2N [structure] COOH

        (c) [structure] COOH   (d) [structure] COOH / COOH

18.11 Following are structural formulas for each salt.

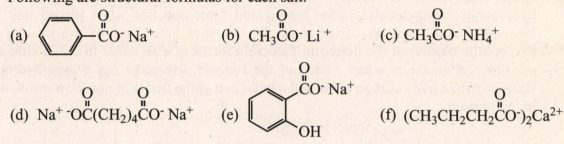

18.13 One of the carboxyl groups in this salt is present as –COO⁻, the other as –COOH.

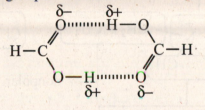

18.15 Hydrogen bonding is between the carbonyl oxygen of one carboxyl group and the hydrogen of the hydroxyl group of the other.

18.17 The polar carboxyl group and the hydrogen bonding of the hydroxyl contribute to water solubility; the hydrocarbon chain prevents water solubility.

18.19 In order of increasing boiling point, they are heptanal (b.p. 153°C), 1-heptanol (b.p. 176°C), and hexanoic acid (b.p. 205°C). The major influence in the high boiling hexanoic acid is the extensive intermolecular hydrogen bonding that exists in carboxylic acids.

18.21 In order of increasing solubility in water, they are decanoic acid, pentanoic acid, and acetic acid. As the hydrocarbon portion of the carboxylic acid molecule decreases, the molecule becomes less hydrophobic and the water solubility increases.

18.23 The structural features necessary to make a good detergent are: (1) a hydrocarbon chain of 12-20 carbon atoms and (2) a polar end group that will not form insoluble salts with Ca^{2+}, Mg^{2+}, or Fe^{3+} ions. The most common detergents are sodium salts of benzenesulfonic acids ($R\text{-}ArSO_3^- Na^+$) where Ar = benzene ring and the -R group is a long hydrocarbon chain at least 12 carbons long.

18.25 Soaps and detergents contain both hydrophilic and hydrophobic components. The hydrophobic components cluster inward and form micelles when placed into water. Nonpolar grease and dirt dissolve in the nonpolar inner part of the micelle. The hydrophilic portion of the detergent favorably interacts with water to aid in emulsifying the dirt and grease in water. Each of the cationic detergents has a long hydrophobic sixteen-carbon hydrocarbon chain and a polar end group that will not form insoluble salts in hard water.

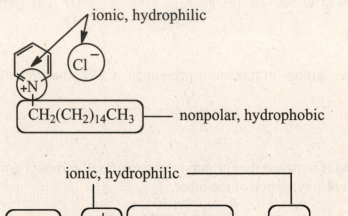

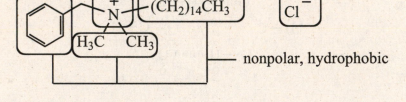

18.27 Carboxylic acids are the stronger acids, with pK_a values in the range 4.0 - 5.0. Phenols are the weaker acids, with pK_a values of approximately 10.0. Alcohols are the weakest of the three, with pK_a values between 16 and 18.

18.29 Following are equations for these acid-base reactions.

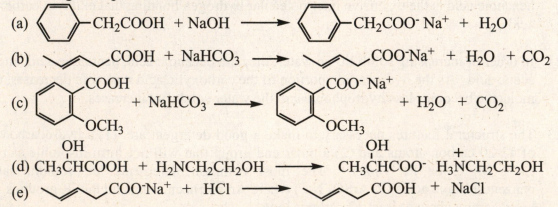

18.31 The acid-base reaction is the neutralization of formic acid by sodium bicarbonate.

$$HCOOH + NaHCO_3 \longrightarrow HCOO^-Na^+ + H_2O + CO_2$$

Formic acid   Sodium formate

18.33 Using $\dfrac{K_a}{[H_3O^+]} = \dfrac{[A^-]}{[HA]}$ and substituting the following values: $\dfrac{[A^-]}{[HA]} = \dfrac{10^{-5}}{[10^{-pH}]}$;

the following table can be generated:

pH	$\dfrac{[A^-]}{[HA]}$
(a) 2.0	1×10^{-3}
(b) 5.0	1×10^{0}
(c) 7.0	1×10^{2}
(d) 9.0	1×10^{4}
(e) 11.0	1×10^{6}

18.35 The pK_a of lactic acid is 4.07. At this pH, lactic acid would be present as 50% $CH_3CH(OH)COOH)$ and 50% $CH_3CH(OH)COO^-$. At pH 7.35 to 7.45, which is more basic than pH 4.07, lactic acid would be present as the lactate anion, $CH_3CH(OH)COO^-$.

18.37 Recall from Chapter 8 that carboxylic acids are stronger acids than ammonium ion. Therefore in part (a), the –COOH group is a stronger acid than the $-NH_3^+$ group.

(a) $CH_3\overset{|}{\underset{NH_3^+}{CH}}COOH + NaOH \longrightarrow CH_3\overset{|}{\underset{NH_3^+}{CH}}COO^-Na^+ + H_2O$

(b) $CH_3\overset{|}{\underset{NH_3^+}{CH}}COO^-Na^+ + NaOH \longrightarrow CH_3\overset{|}{\underset{NH_2}{CH}}COO^-Na^+ + H_2O$

18.39 In part (a), the amino group (-NH₂) is a stronger base than the carboxylate group (-COO⁻).

(a) $CH_3\overset{|}{\underset{NH_2}{CH}}COO^-Na^+ + HCl \longrightarrow CH_3\overset{|}{\underset{NH_3^+}{CH}}COO^-Na^+ + Cl^-$

(b) $CH_3\overset{|}{\underset{NH_3^+}{CH}}COO^-Na^+ + HCl \longrightarrow CH_3\overset{|}{\underset{NH_3^+}{CH}}COOH + NaCl$

18.41 Following is a structural formula for the ester formed in each reaction. Water is also formed along with each indicated ester.

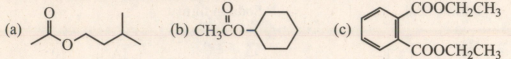

(a)   (b) CH$_3$CO—   (c)

18.43 Following is a structural formula for methyl 2-hydroxybenzoate.

18.45 Phenylacetic acid is unreactive toward NaBH$_4$ in (e) and Ni/H$_2$ in (g).
(a) C$_6$H$_5$CH$_2$COO$^-$Na$^+$ + CO$_2$ + H$_2$O   (b) C$_6$H$_5$CH$_2$COO$^-$Na$^+$ + H$_2$O
(c) C$_6$H$_5$CH$_2$COO$^-$NH$_4^+$ (d) C$_6$H$_5$CH$_2$CH$_2$OH
(f) C$_6$H$_5$CH$_2$CO$_2$CH$_3$ + H$_2$O

18.47 Each synthesis involves a Fischer esterification between p-aminobenzoic acid and the desired alcohol. Fischer esterifications require an acid catalyst, but the p-aminobenzoic acid has a basic amino group that would consume an acid catalyst, therefore, p-aminobenzoic acid must first be converted the ammonium salt. H$_2$SO$_4$ is added to completely protonate the amino acid and to catalyze the esterification with the alcohol. Then, dilute Na$_2$CO$_3$ is added to convert the ammonium salt back to the amine.

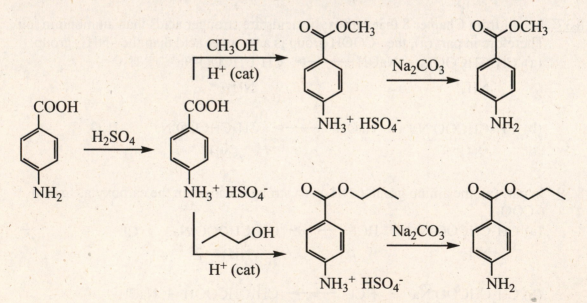

18.49 The ester is a six-membered lactone.

18.51 The following starting materials are oxidized to the indicated products.

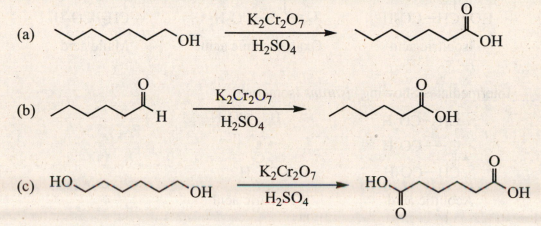

18.53 (a) TCA intermediates that are chiral and those that show *cis/trans* isomerism are shown below. The stereocenters are identified with an asterisk. Isocitric acid has the most chiral centers.

Chiral Intermediates:

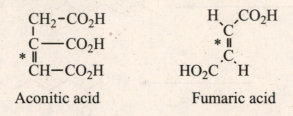

Isocitric acid   Oxalosuccinic acid   Malic acid

Intermediates Showing *cis/trans* Isomerism

Aconitic acid   Fumaric acid

(b) Step 1 is dehydration of a 3° alcohol to an alkene, Step 2 is a hydration of an alkene to a 2° alcohol, and Step 3 is an oxidation of a 2° alcohol to a ketone.

(c) A Markovnikov addition of water to aconitic acid produces citric acid. The fact that the hydration proceeds as it does is due to the presence of an enzyme. This enzyme positions the aconitic acid at the reactive site on the enzyme surface in such a way that only one carbocation intermediate can be formed and water then adds regiospecifically.

(d) Step 4 is a decarboxylation of a β-ketoacid, Step 5 is a decarboxylation with an oxidation (oxidative decarboxylation), Step 6 is an oxidation (dehydrogenation), Step 7 is a hydration of a carbon-carbon double bond to a 2° alcohol, and Step 8 is an oxidation of a 2° alcohol to a ketone.

(e) The oxidative decarboxylation can be described as the loss of CO_2 to form an aldehyde, then the aldehyde oxidized to a carboxylic acid.

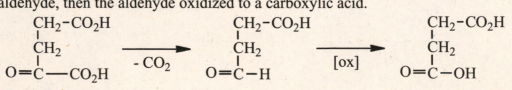

α-Ketoglutaric acid

(f) This decarboxylation is simply the loss of CO_2 with no overall oxidation/reduction taking place because no new oxygen is added.

Chapter 19: Carboxylic Anhydrides, Esters, and Amides

19.1 The following is a structural formula for each amide.

(a) cyclohexyl-NHCCH₃ ... (b) phenyl-CNH₂

19.2 Under basic conditions, as in part (a), each carboxyl group is present as a carboxylic anion. Under acidic conditions, as in part (b), each carboxyl group is present in its un-ionized form.

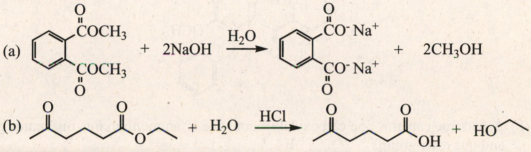

(a) [benzene with two COCH₃ groups] + 2NaOH $\xrightarrow{H_2O}$ [benzene with two CO⁻Na⁺ groups] + 2CH₃OH

(b) [diketone ester] + H₂O $\xrightarrow{HCl}$ [diketo acid with OH] + HO—CH₂CH₃

19.3 In aqueous NaOH, each carboxyl group is present as a carboxylic anion, and each amine is present in its unprotonated form.

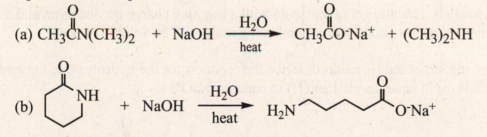

(a) CH₃CN(CH₃)₂ + NaOH $\xrightarrow[\text{heat}]{H_2O}$ CH₃CO⁻Na⁺ + (CH₃)₂NH

(b) [cyclic amide (piperidinone), NH] + NaOH $\xrightarrow[\text{heat}]{H_2O}$ H₂N—(CH₂)₄—CO⁻Na⁺

19.5 (a) Benzoic anhydride   (b) Methyl decanoate   (c) *N*-Methylhexanamide
(d) 4-Aminobenzamide   (e) Cyclopentyl acetate   (f) Ethyl 3-hydroxybutanoate
   or *p*-Aminobenzamide   or Cyclopentyl ethanoate

19.7 Reaction (a) is saponification. Reaction (b) is an acid-catalyzed ester hydrolysis.

(a) [phenyl]—COCH₂CH₃ + NaOH $\xrightarrow{H_2O}$ [phenyl]—CO⁻Na⁺ + CH₃CH₂OH

(b) [phenyl]—COCH₂CH₃ + H₂O $\xrightarrow{HCl}$ [phenyl]—COH + CH₃CH₂OH

19.9 Only (a), a carboxylic acid has an acidic proton and will react with sodium bicarbonate to produce carbonic acid, which then decomposes to water and carbon dioxide.

19.11 Following is a structural formula for phenacetin.

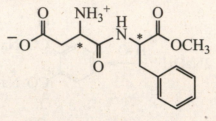

$$CH_3CH_2O--NH-\overset{\overset{O}{\|}}{C}CH_3$$

19.13 (a) Aspartame has two stereocenters and is chiral. Four stereoisomers (two pairs of enantiomers) are possible.

(b) Aspartame contains one carboxylate anion, one 1° ammonium ion, one amide group, and one ester group.

(c) Aspartame will have a net charge of zero at pH of 7.

(d) Aspartame is a zwitterion (a molecule with a negative charge on one atom and a positive charge on another atom) and being ionic, will be soluble in water.

The following structural formulas describe the products for the hydrolysis of the ester and amide bonds (e) in aqueous HCl and (f) in aqueous NaOH .

(e) Hydrolysis in HCl:

(f) Hydrolysis in NaOH:

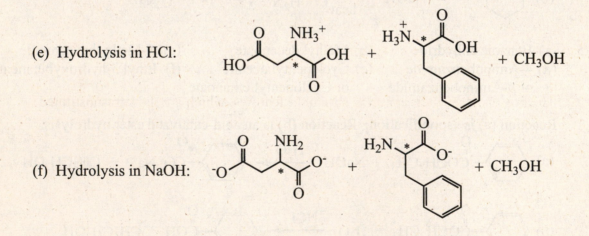

19.15 Following are sections of two parallel nylon-66 chains, with hydrogen bonds between N-H and C=O groups indicated by dashed lines.

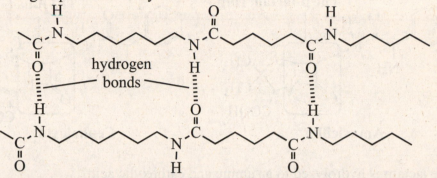

19.17 In anhydrides of carboxylic acids, the functional group is two carbonyl (C=O) groups bonded to an oxygen atom. In an anhydride of phosphoric acid, the functional group is two phosphoryl (P=O) groups bonded to an oxygen atom.

19.19 The following is the structural formula of dihydroxyacetone phosphate shown as it would be when ionized at pH 7.40.

$$\overset{\quad\;\;O}{\underset{\underset{O^-}{|}}{O{-}P}}{-}OCH_2\overset{O}{\overset{||}{C}}CH_2OH$$

19.21 Hydrolysis requires one mole of water per mole of ester.

$$CH_3O{-}\overset{\overset{O}{||}}{\underset{\underset{OCH_3}{|}}{P}}{-}OCH_3 \;+\; H_2O \;\xrightarrow{\text{NaOH}}\; CH_3O{-}\overset{\overset{O}{||}}{\underset{\underset{OCH_3}{|}}{P}}{-}O^-\,Na^+ \;+\; CH_3OH$$

Trimethyl phosphate   Dimethyl phosphate

19.23 The box encloses their common structural features, which are the tetrasubstituted cyclopropane ring, the ester group, the trisubstituted carbon double bond on the cyclopropane ring, and the *trans* configuration of the ester group and the trisubstituted double bond.

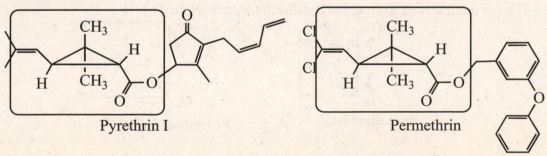

Pyrethrin I   Permethrin

19.25 (a) The β-lactam ring is shown in color in Chemical Connections 19B.

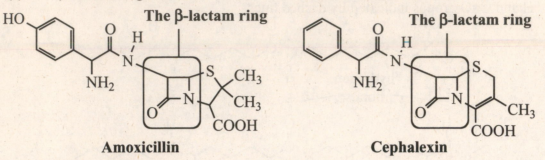

Amoxicillin   **Cephalexin**

(b) The lactam is hydrolyzed to an amine and carboxylic acid.

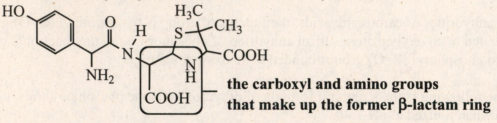

the carboxyl and amino groups
that make up the former β-lactam ring

19.27 The aspirin molecule contains a carboxyl group and an ester group.

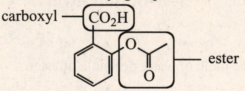

19.29 Both aspirin and ibuprofen contain a carboxylic acid and a benzene ring. Naproxen also contains a carboxylic acid and a benzene ring; in naproxen, the benzene ring is a part of a naphthalene ring.

19.31 A sunblock prevents all ultraviolet radiation from reaching protected skin by reflecting it away from the skin. A sunscreen absorbs a portion of the ultraviolet radiation hitting protected skin and radiates it as heat.

19.33 The portion derived from urea contains the atoms –NH-CO-NH–.

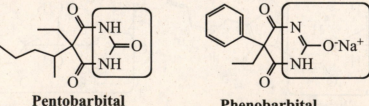

Pentobarbital   **Phenobarbital**

19.35 Following is a structural formula for benzocaine, ethyl 4-aminobenzoate.

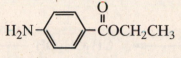

19.37 Hydrolysis gives one mole of 2,3-dihydroxypropanoic acid and two moles of phosphoric acid. At pH 7.35 - 7.45, the carboxyl group is present as its anion, and phosphoric acid is present as its dianion.

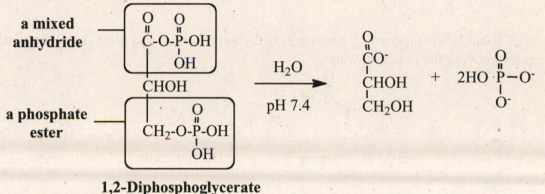

1,2-Diphosphoglycerate

19.39 Following is a short section of the polyamide formed in this type of polymerization. In the formation of the polymer, a molecule of water is lost for every amide bond formed.

reactive toward amine end of an amino acid

$$H-N-CHC-OH + H-N-CHC-OH + H-N-CHC-OH \xrightarrow{-H_2O}$$

reactive toward carboxylic acid end of an amino acid

Polyamide

19.41 Using ethylene as the only carbon source, ethyl acetate can be synthesized using the following reactions:

$$H_2C = CH_2 + H_2O \xrightarrow{H_2SO_4} CH_3CH_2OH \xrightarrow[H_2SO_4]{K_2Cr_2O_7} CH_3CO_2H$$

$$CH_3CO_2H + CH_3CH_2OH \xrightarrow{H_2SO_4} CH_3\overset{\overset{\displaystyle O}{\|}}{C}OCH_2CH_3 + H_2O$$

19.43 DEET can be synthesized by heating 3-methylbenzoic acid and diethylamine together and removing the resulting water.

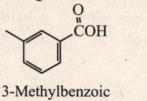

3-Methylbenzoic
acid

Diethylamine

19.45 Barbital has the following structure:

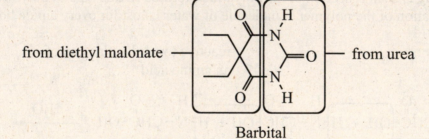

from diethyl malonate —— —— from urea

Barbital

19.47 (a) The conformation on the left is the most stable. Placing three large groups in the equatorial position instead of in the less favorable axial positions minimizes the 1,3-diaxial repulsions.

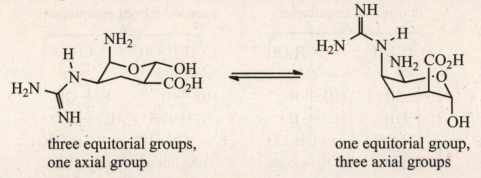

three equitorial groups,
one axial group

one equitorial group,
three axial groups

(b) The hydroxyl oxygen (O6) reacts with aldehyde carbonyl carbon (C6) in **B** to form a cyclic hemiacetal, **A**.

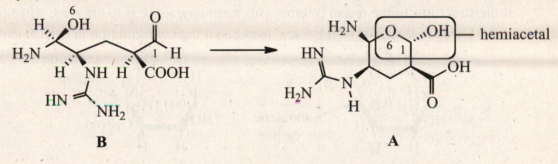

B

A

Chapter 20: Carbohydrates

20.1 Following are Fischer projections for the four 2-ketopentoses. They consist of two pairs of enantiomers.

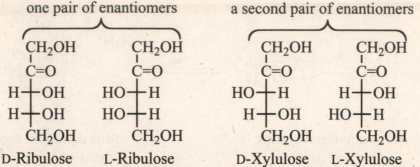

one pair of enantiomers   a second pair of enantiomers

D-Ribulose   L-Ribulose   D-Xylulose   L-Xylulose

20.2 D-Mannose differs in configuration from D-glucose only at carbon 2. One way to arrive at the structures of the α and β forms of D-mannopyranose is to draw the corresponding α and β forms of D-glucopyranose, and then invert the configuration in each at carbon 2.

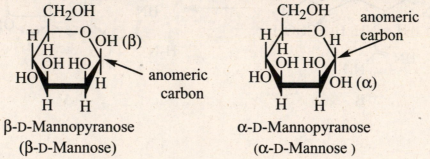

β-D-Mannopyranose   α-D-Mannopyranose
(β-D-Mannose)   (α-D-Mannose)

20.3 D-Mannose differs in configuration from D-glucose only at carbon 2.

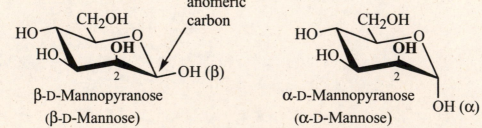

β-D-Mannopyranose   α-D-Mannopyranose
(β-D-Mannose)   (α-D-Mannose)

20.4 Following is a Haworth projection and a chair conformation for methyl α- D-mannopyranoside.

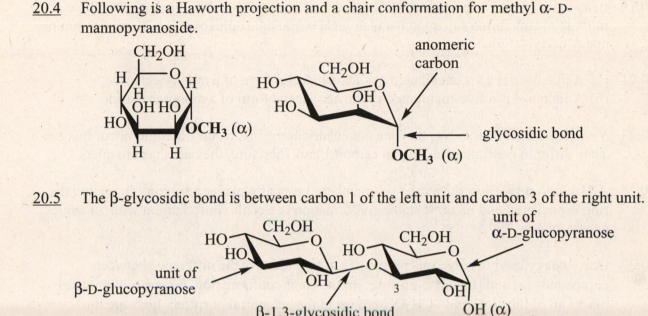

20.5 The β-glycosidic bond is between carbon 1 of the left unit and carbon 3 of the right unit.

20.7 The carbonyl group in an aldose is an aldehyde. In a ketose, the carbonyl is a ketone. An aldopentose is an aldose with five carbons. A ketopentose is a ketose with five carbons.

20.9 The three most abundant hexoses are D-glucose, D-galactose, and D-fructose. The first two are aldohexoses. The third is a 2-ketohexose.

20.11 To say that they are enantiomers means that they are nonsuperposable mirror images.

20.13 The D or L configuration in an aldopentose is determined by its configuration at carbon 4.

20.15 Compounds (a) and (c) are D-monosaccharides. Compound (b) is an L-monosaccharide.

20.17 A 2-ketoheptose has four stereocenters and 16 possible stereoisomers. Eight of these are D-2-ketoheptoses and eight are L-2-ketoheptoses. Following is one of the eight possible D-2-ketoheptoses.

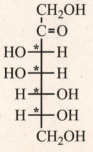

20.19   In an amino sugar, one or more –OH groups are replaced by $-NH_2$ groups. The three most abundant amino sugars in the biological world are D-glucosamine, D-galactosamine, and N-acetyl- D-glucosamine.

20.21   (a) A pyranose is a six-membered cyclic hemiacetal form of a monosaccharide.
        (b) A furanose is a five-membered cyclic hemiacetal form of a monosaccharide.

20.23   Yes, they are anomers. No, they are not enantiomers; that is, they are not mirror images. They differ in configuration only at carbon 1 and, therefore, they are diastereomers.

20.25   A Haworth projection shows the six-membered ring as a planar hexagon. In reality, the ring is puckered and its most stable conformation is a chair conformation with all angles approximately 109.5°.

20.27   One strategy used to solve this problem is to first identify the difference between compounds (a) and (b) and D-glucose. In the chair conformation, β-D-glucopyranose places all of the –OH and $-CH_2OH$ groups in the equatorial position. Ignoring the configuration at the anomeric carbon (C1) in the cyclic form, compound (a) differs from D-glucose only in the configuration at carbon 4. Compound (b) differs only at carbon 3.

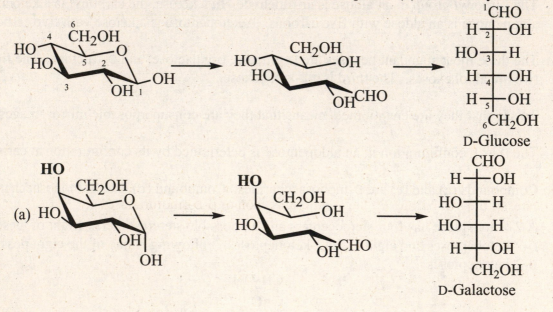

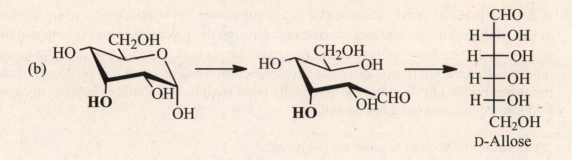

(b)

D-Allose

20.29 The specific rotation of α-L-glucose is -112.2°.

20.31 A glycoside is a cyclic acetal of a monosaccharide. A glycosidic bond is the bond from the anomeric carbon to the –OR group of the glycoside.

20.33 No, glycosides cannot undergo mutarotation because the anomeric carbon is not free to interconvert between α and β configurations via the open-chain aldehyde or ketone.

20.35 Following are Fischer projections of D-glucose and D-sorbitol. The configurations at the four stereocenters of D-glucose are not affected by this reduction.

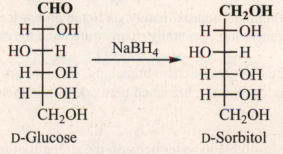

D-Glucose   D-Sorbitol

20.37 Ribitol is the reduction product of D-Ribose. β-D-Ribose 1-phosphate is the phosphoric ester of the –OH group on the anomeric carbon of β-D-ribofuranose.

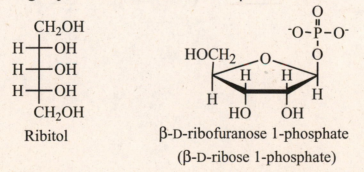

Ribitol   β-D-ribofuranose 1-phosphate
(β-D-ribose 1-phosphate)

20.39 A β-1,4-glycosidic bond indicates that the configuration at the anomeric carbon (carbon 1 in this problem) of the monosaccharide unit forming the glycosidic bond is beta and that it is bonded to carbon 4 of the second monosaccharide unit. An α-1,6-glycosidic bond indicates that the configuration at the anomeric carbon (carbon 1 in this problem) of the monosaccharide unit forming the glycosidic bond is alpha and that it is bonded to carbon 6 of the second monosaccharide unit.

20.41 (a) Both monosaccharide units are D-glucose.
(b) They are joined by a β-1,4-glycosidic bond.
(c) It is a reducing sugar.
(d) It undergoes mutarotation.

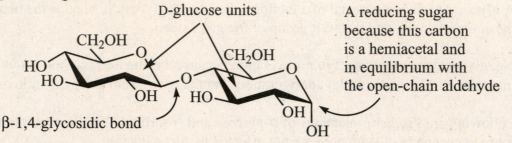

D-glucose units

A reducing sugar because this carbon is a hemiacetal and in equilibrium with the open-chain aldehyde

β-1,4-glycosidic bond

20.43 An oligosaccharide contains approximately six to ten monosaccharide units. A polysaccharide contains more, generally many more, than ten monosaccharide units.

20.45 The difference is in the degree of chain branching. Amylose is composed of unbranched chains, whereas amylopectin is a branched network with branches started by α-1,6-glycosidic bonds.

20.47 Cellulose fibers are insoluble in water because the strength of hydrogen bonding of a cellulose molecule in the fiber with surface water molecules is not sufficient to overcome the intermolecular forces that hold it in the fiber.

20.49 (a) In the following structural formula, the $CH_3C{=}O$ (the acetyl group) is abbreviated Ac.

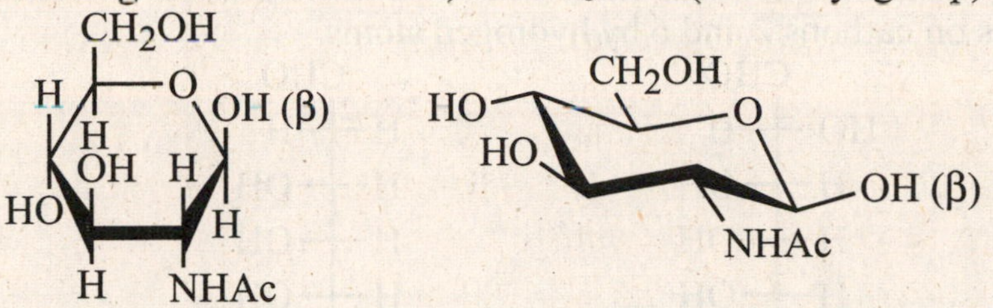

(b) The following Haworth and chair structures for this repeating disaccharide.

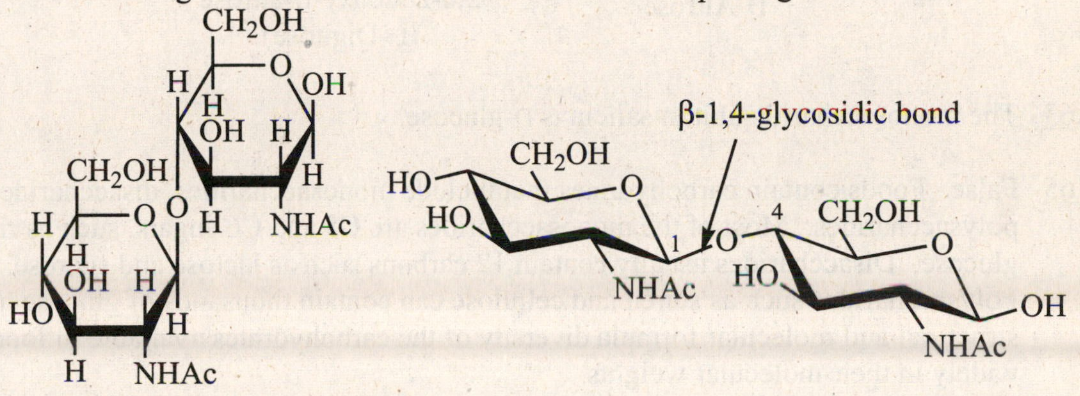

20.51 Its lubricating power decreases.

20.53 Sucrose can be substituted for lactose. Sucrose is a disaccharide composed of D-glucose and D-fructose, thus lacks the problematic D-galactose. With maturation, children develop an enzyme capable of metabolizing D-galactose; therefore, they are able to tolerate galactose as they get older.

20.55 L-Ascorbic acid is oxidized (there is loss of two hydrogen atoms) when it is converted to L-dehydroascorbic acid. L-Ascorbic acid is a biological reducing agent.

20.57 Types A, B and O have in common D-galactose and L-fucose. Only type A has N-acetyl-D-glucosamine.

20.59 Mixing types A and B blood will result in coagulation.

20.61 Consult Table 20.1 for the structural formula of D-altrose and draw it. Then replace the -
—OH groups on carbons 2 and 6 by hydrogen atoms.

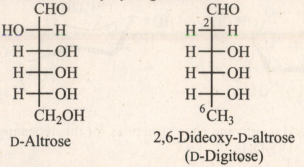

D-Altrose

2,6-Dideoxy-D-altrose
(D-Digitose)

20.63 The monosaccharide unit in salicin is D-glucose.

20.65 False. Foods contain carbohydrates that include monosaccharides, disaccharides, and polysaccharides. Most of the monosaccharides are C5 and C6 sugars, such as ribose and glucose. Disaccharides usually contain 12 carbons such as lactose and sucrose. Polysaccharides such as starch and cellulose can contain thousands of glucose units. The structural and molecular formula diversity of the carbohydrates available in foods vary widely in their molecular weights.

20.67 The Haworth projection of the fructose moiety in sucrose is the most accurate representation of its nearly planar structure.

20.69 In starch, α-glycosidic bonds join one glucose unit to another. Cellulose has β-glycosidic bonds joining glucose units to each other. This difference means that enzymes present in humans and other animals are specific to cleaving the α-glycosidic bonds, aiding in the digestion of starch but not cellulose. Structurally, the hydroxyl groups in cellulose are involved in extensive intermolecular hydrogen bonding between cellulose chains, leaving no opportunity for hydrogen bonding with water, rendering cellulose insoluble in water. The α-glycosidic bonds in a more water-soluble starch give a structure that directs the hydroxyls outward, enabling hydrogen bonding with water.

20.71 The ring system on the upper left is a sugar (glucose). The presence of the cyanide group is he main cause of concern. The metabolism of laetrile produces hydrogen cyanide (HCN).

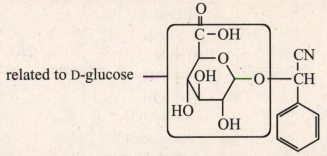

related to D-glucose

20.73 The only structural difference between D-glucose and D-galactose is the stereochemistry at C4. They are epimers (stereoisomers of two or more chiral centers that differ by only one chiral center), which belong to the broader class of diastereomers. The conversion of D-galactose to D-glucose occurs through the Leloir Pathway, which uses the enzymes galactokinase, galactose-1-phosphate uridyltransferase, and UDP-galactose-4'-epimerase. This conversion can be considered an epimerization, in that it only inverts one chiral center between the two molecules.

$$^{1}CHO \qquad CHO$$

	D-Glucose		D-Galactose
H—OH (2)		H—OH	
HO—H (3)		HO—H	
H—OH (4)		HO—H	
H—OH (5)		H—OH	
$_6CH_2OH$		CH_2OH	

D-Glucose   D-Galactose

20.75 Chitin is a polymer of repeating *N*-acetyl-D-glucosamine units.

1,4-β-glycosidic bond

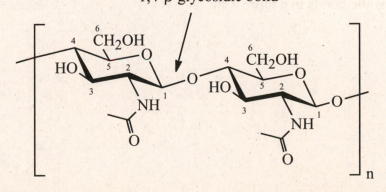

20.77 Review Section 17.5 on keto-enol tautomerism and your answer to Problem 17.71. The intermediate in this conversion is an enediol; that is, it contains a carbon-carbon double bond with two –OH groups on it.

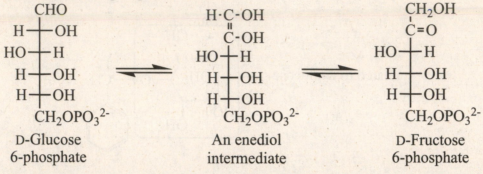

D-Glucose 6-phosphate   An enediol intermediate   D-Fructose 6-phosphate

20.79 (a) The left unit is D-glucuronic acid, which is derived from D-glucose. The right unit is a sulfate ester derived from N-acetyl- D-galactosamine, which is in turn derived from D-galactosamine.

   (b) The two units are joined by a β-1,3-glycosidic bond.

20.81 Cellulose molecules are held together by intermolecular hydrogen bonding. The addition of water disrupts this hydrogen bonding and compromises the strength of paper. Oil is nonpolar and does not interact with the cellulose hydroxyls in paper, therefore does not affect the structural integrity of paper.

21.1 (a) This complex lipid is produced from the triol glycerol. Two of the hydroxyl groups are esterified with fatty acids, the third with phosphoric acid. Therefore, the lipid is broadly classified as a glycerophospholipid. The phosphate group is esterified with the hydroxyl group of the amino acid serine; therefore, the lipid belongs to the subgroup cephalins.
(b) The components present: glycerol backbone, myristic acid, linoleic acid, phosphate, and serine.

21.3 Lipids are defined in terms of solubility rather than by a defined chemical structure. Lipids are insoluble in water and soluble in low polarity solvents like diethyl ether.

21.5 The melting points of fatty acids (Tables 18.3 and 21.1) are dependent on the length of the carbon chains and the number and types of double bonds. Cis double bonds cause a kink in the chain (see Section 21.2) that disrupts the London dispersion forces between the chains and lowers the melting point of oleic acid (18:1) compared to the saturated acid, stearic acid (18:0). However, the trans fatty acid (18:1), resembles a saturated acid and does not disrupt the chain as much as the cis acid. Thus the melting point would be predicted to be higher. Actual melting points: 18:0 (70° C); 18:1 trans (45° C); 18:1 cis (16° C).

21.7 The diglycerides with the highest melting points will be those with two stearic acids (a saturated fatty acid). The lowest melting point will be the one with two oleic acids (a monounsaturated fatty acid).

21.9 A triglyceride containing only stearic acid (18:0) will have a higher melting point than a triglyceride containing only lauric acid (12:0). The correct answer is (b).

21.11 The highest melting triglyceride would be (a), containing palmitic and stearic (both saturated fatty acids). The lowest melting triglyceride is (c) because all fatty acids are unsaturated. Triglyceride (b), which has one unsaturated fatty acid and two saturated acids, would have a melting point in between triglycerides (a) and (c). Therefore, the order from lowest to highest is (c), (b), (a)

21.13 In order of increasing solubility in water, they are (a) < (b) < (c). Solubility in water is dependent on the presence of polar functional groups like the hydroxyl group. The greater the number of hydroxyl groups, the greater the solubility. A monoglyceride has two hydroxyl groups, a diglyceride has one hydroxyl group, and a triglyceride has none. Fatty acids are essentially insoluble in water.

21.15 Saponification of the triglyceride hydrolyzes the ester bonds yielding glycerol and the sodium salts of the fatty acids palmitic, stearic, and linolenic.

21.17 Complex lipids can be divided into two groups: phospholipids and glycolipids. Phospholipids contain an alcohol, two fatty acids, and a phosphate group. There are two types: glycerophospholipids and sphingolipids. In glycerophospholipids, the alcohol is glycerol. In sphingolipids, the alcohol is sphingosine. Glycolipids are complex lipids that contain carbohydrates.

20.19 The fluidity of a membrane is dependent on the amount of cholesterol and on the amounts and types of fatty acids in the phospholipid bilayer. The presence of cis double bonds causes greater fluidity because they cannot pack together as closely as saturated fatty acids. The greater the concentration of lipids with unsaturated fatty acids, the greater the fluidity of the membrane.

21.21 Integral membrane proteins are embedded within the membrane and are held in position by very strong hydrophobic interactions. They often span the entire lipid bilayer where they serve as channels through which molecules are transported in and out of a cell. Peripheral membrane proteins are present on the surfaces of the membrane where they are weakly attached.

21.23 A phosphatidylinositol containing oleic acid and arachidonic acid:

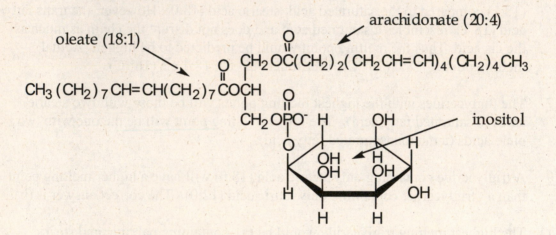

21.25 Complex lipids that contain ceramides include sphingomyelin, a sphingolipid, and the cerebroside glycolipids.

21.27 The hydrophilic functional groups of
(a) glucocerebroside: carbohydrate; hydroxyl and amide group of the ceramide.
(b) sphingomyelin: phosphate group; choline; hydroxyl and amide of ceramide.

21.29 Cholesterol crystals may be found in (1) gallstones, that are sometimes pure cholesterol; (2) joints of people suffering from bursitis.

21.31 Carbon 17 of the D steroid ring undergoes the most substitution.

21.33 LDL in the bloodstream is delivered to cells by binding to LDL receptor proteins in areas called coated pits on the surfaces of the cells. After binding, the LDL is transported inside the cells (endocytosis) where the cholesterol is released by enzymatic degradation of the LDL.

21.35 VLDL particles, which have mainly a triglyceride core, are produced in the liver. They enter the blood and are carried to fat and muscle tissue where they deliver triglycerides for energy metabolism. Removing the lipid from the VLDL core increases their density and converts them to LDL particles.

21.37 Serum cholesterol levels control the production of LDL receptors that help cells take up cholesterol in the form of LDL. When serum cholesterol concentration is high, the synthesis of cholesterol in the liver is inhibited and the synthesis of cell receptors is increased. Serum cholesterol levels control the formation of cholesterol in the liver by regulating enzymes that synthesize cholesterol. This is an example of feed-back control.

21.39 Estradiol (E) is synthesized from progesterone (P) through the intermediate testosterone (T). First the D-ring acetyl group of P is converted to a hydroxyl group and T is produced. The methyl group in T, at the junction of rings A and B, is removed and ring A becomes aromatic. The keto group in P and T is converted to a hydroxyl group in E.

21.41 The structures of the three steroids are shown in Figure 21.7. They have similar substituents at several positions, but differ greatly at C-11. Progesterone has no substituents except hydrogen, cortisol has a hydroxyl group, cortisone has a keto group, and RU486 has a large p-aminophenyl group. Apparently the functional group at C-11 is of little importance in binding the compounds to the receptor.

21.43 Common structural features in oral contraceptives: they all are based on the four-fused steroid ring; most have structures similar to progesterone; all have a methyl group at C-13; all have an acetylenic group on C-17; all have some unsaturation in ring A and/or B.

21.45 The bile salts help solubilize fats during the digestion process and elimination of waste products. They assist in removal of excess cholesterol that causes plaque formation in atherosclerosis in two ways: (1) They themselves are oxidation products of cholesterol degradation so this decreases cholesterol concentrations, and (2) they bind to cholesterol and form complexes that are eliminated in the feces.

21.47 (a) Glycocholate:

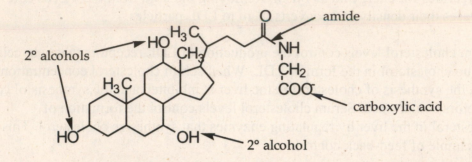

(b) Cortisone:

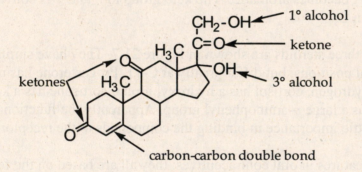

(c) PGE₂:

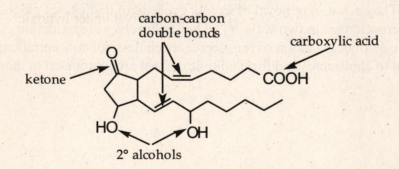

(d) Leukotriene B4:

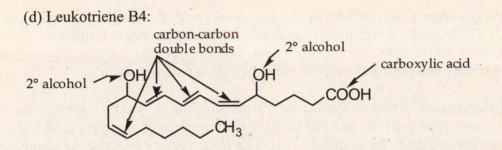

21.49 Thromboxanes, which are derived from arachidonic acid via the prostaglandins, stimulate platelet aggregation thus enhancing the blood clotting process. Aspirin slows the synthesis of the thromboxanes by inhibiting the COX enzyme. Thus in the presence of aspirin, there will be lowered amounts of thromboxanes and the blood clotting process will be slowed. Strokes caused by blood clots in the brain will occur less often.

21.51 The major components of waxes are esters of long chain acids and alcohols. Beeswax contains palmitic acid (16:0) and 1-triacontanol, a saturated, unbranched, primary alcohol with thirty carbons. Because of the presence of saturated components, wax molecules pack closer together in the solid form and have higher melting points than triglycerides that often contain many unsaturated components.

21.53 The anion transporter of the erythrocyte membrane exchanges chloride and bicarbonate ions. The protein transporter forms a hydrophilic channel through the membrane. Protein polar groups in the channel wall facilitate the transport of hydrated chloride and bicarbonate ions. The hydrophobic groups on the outer surfaces of the helices interact with the membrane.

21.55 The proposed link between ceramide length and response to hypoxia comes from studies in the roundworm *C. elegans*. Experiments showed a connection between expression of short chain ceramides (C_{20-22}) and cell survival under hypoxic conditions. Conversely, when long chain ceramides (C_{24-26}) were produced, the cells died under hypoxic conditions. This finding may have important ramifications for human health for a number of reasons. First, many genes from *C. elegans* are also found in humans so the link between ceramide length and hypoxia may hold true. Secondly, inhibitors can be designed to inhibit specific ceramide synthases thereby improving stroke and heart attack survivability. Lastly, similar drugs could be used to make cancer cells more susceptible to cell death caused by hypoxia by increasing ceramide chain length.

21.57 The glycolipid that accumulates in Fabry's disease contains the monosaccharides galactose and glucose.

21.59   Athletes use steroids to increase muscle mass, especially important in power sports and in low doses to also to aid in recovery after athletic events.

21.61   Progesterone is secreted during the menstrual cycle after ovulation (see Figure 21.8). Its purpose is to prevent another ovulation if the egg is fertilized. Contraceptive agents give the user a level of progesterone that sends the message that there is a fertilized egg; therefore, ovulation does not occur.

21.63   Indomethacin, a non-steroidal anti-inflammatory agent, acts by inhibiting the cyclo- oxygenase enzymes that catalyze the formation of prostaglandins from arachidonic acid. Prostaglandins cause an inflammatory response so inhibiting their formation will reduce inflammation.

21.65   The leukotrienes are synthesized from arachidonic acid as are prostaglandins, but not through the same pathway using the COX enzymes. Thus, the NSAIDS inhibit the cyclo- oxygenases and prostaglandin synthesis, but do not effect leukotriene synthesis. Aspirin and other NSAIDS agents can reduce inflammation, but they do not alter the physiological actions of the leukotrienes.

21.67   Omega-3 fats inhibit the synthesis of certain prostaglandins and thromboxane A – a molecule similar to prostaglandins. Thromboxane A is released by ruptured arteries and helps platelets aggregate and form blood clots. Thus, by eating a diet rich in fish there is a lower tendency to form blood clots that can cause heart attacks and strokes.

21.69   (Study membrane structure in Figure 21.2). Three regions of the membrane need to be considered in order to define transport: 1) the cytoplasmic surfaces and 2) the exterior surfaces that are comprised primarily of the polar heads of complex lipids (purple balls) and 3) the membrane interior composed of the hydrophobic tails of the lipids. To diffuse, molecules must pass through all three regions, two polar regions and one nonpolar region. Polar molecules interact favorably with the surface regions by hydrogen bonding and ion-dipole interactions, but they repel the hydrophobic barrier, thus they would not diffuse through the membrane. In addition, polar molecules would prefer to interact with water that is present at each surface. On the other hand, small, nonpolar molecules are able to diffuse because they slip through surface regions containing mainly hydrophobic components (lipid patches).

21.71   Although prostaglandins and thromboxanes have different structures, the physiological production of both is slowed by COX inhibitors. This suggests that there must be a common step in their syntheses. Prostaglandins are synthesized from arachidonic acid using the COX enzymes. Thromboxanes are synthesized from the prostaglandin PGH.

21.73   Coated pits are areas on a cell membrane that have high concentrations of LDL receptors and participate in transporting LDLs into the cell.

21.75   In facilitated transport, a membrane protein assists in the movement of a molecule through the membrane. The protein often becomes a channel for transport of the molecule. No energy is required for the transport process. In the active transport of a molecule, a membrane protein assists in the process, but energy, usually released from ATP hydrolysis, is required.

21.77   Aldosterone has an aldehyde group at carbon 13.

21.79   The formula weight of the triglyceride is about 850 g/mole. 100 g of the triglyceride is 100 g ÷ 850 g/mole = 0.12 mole. One mole of hydrogen is required for each mole of double bonds in the triglyceride. There are three double bonds so the moles of hydrogen required for 100 g = 0.12 mole x 3 = 0.36 mole of hydrogen gas. Converting to grams of hydrogen, 0.36 x 2 g/mole = 0.72 g hydrogen gas.

21.81   By definition, a molecule comprising a fatty acid and sphingosine is a ceramide.

21.83   No. There are many proteins that are found imbedded in the membrane but protruding through one side only. Others are loosly attached to one side of the membrane.

21.85   Statements (c) and (d) are consistent with what is known about membranes. Covalent bonding between lipids and proteins [statement (e)] is not widespread. Proteins "float" in the lipid bilayers rather than being sandwiched between them [statement (a)]. Bulkier molecules tend to be found in the outer lipid layer [statement (b)].

21.87   Statement (c) is correct. Transverse diffusion is only rarely observed statement (b)]. Proteins are bound to the inside and outside of the membrane [statement (a)].

21.89   Both lipids and carbohydrates contain carbon, hydrogen, and oxygen. Carbohydrates have aldehyde and ketone groups, as do some steroids. Carbohydrates have a number of hydroxyl groups, which lipids do not have to a great extent. Lipids have major components that are hydrocarbon in nature. These structural features imply that carbohydrates tend to be significantly more polar than lipids.

21.91   Primarily lipid: olive oil and butter; primarily carbohydrate: cotton and cotton candy.

21.93   The amounts are the key point here. Large amounts of sugar can provide energy. Fat burning due to the presence of taurine plays a relatively minor role because of the small amount.

21.95   The other end of the molecules involved in the ester linkages in lipids, such as fatty acids, tend not to form long chains of bonds with other molecules. Amino acids have two ends with functional groups, so these groups can link together to form chains.

21.97 Stem cells are progenitor cells for all cell types. They have the ability to develop into any cell type as well as to replicate into more stem cells. In myeloproliferative diseases too many immune cells are produced, i.e. leukemia by the bone marrow stem cells that produce blood cells also known as hematopoietic cells.

21.99 The bulkier molecules will tend to be found on the exterior of the cell because the curvature of the cell membrane will provide more room for them.

21.101 The charges will tend to cluster on membrane surfaces. Positive and negative charges will attract each other. Two positive or two negative charges will repel, so unlike charges do not have this repulsion.

Chapter 22: Proteins

22.1 The dipeptide Val-Phe:

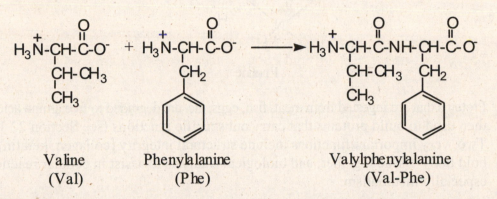

      Valine   Phenylalanine   Valylphenylalanine
      (Val)   (Phe)   (Val-Phe)

22.3 Cobalt is in group 9 of the periodic table so it has 9 valence electrons when the charge is zero. However, $[Co(NH_3)_5Cl]^{2+}$ has a chloride ligand that has a 1-charge and 5 ammonia ligands providing zero charge. Thus, the charge on cobalt must be 3+ for the total charge on the compound to be 2+. That means that cobalt III only has 6 valence electrons. It needs 12 more electrons to achieve the 18 electron rule. Each ammonia ligand provides a pair of electrons through a coordinate covalent bond for a total of 10 electrons and the chloride ligand provides 2 more additional electrons for a grand total of 12 electrons. These 12 electrons on the ligands plus the original 6 valence electrons on Co III satisfy the 18 electron rule.

22.5 (a) Ovalbumin is a protein found in eggs that serves to store nutrients for the developing embryo.
(b) The protein myosin is essential for muscle contraction.

22.7 Immunoglobulins are complex protein structures that serve for protection. Antibodies are immunoglobulins, as are certain receptors on T cells and B cells.

22.9 See Figure 22.1. Tyr has a hydroxyl group on the phenyl ring; Phe has a hydrogen at the same position.

22.11 Arg has the highest percentage of nitrogen at 32%. The four nitrogens have a weight of $4 \times 14 = 56$. The molecular weight of Arg is 174. Percent of N = $56 \div 174 \times 100 = 32\%$. His has 27% nitrogen.

22.13 The amino acid Pro, which has a five-membered ring containing a single nitrogen atom, is classified as a pyrrolidine, a heterocyclic aliphatic amine:

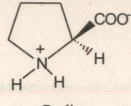

Proline

22.15 Proteins that are ingested from meat, fish, eggs, etc. are degraded to free amino acids and then used to build proteins that carry out specific functions (see Section 22.1). Two very important functions include structural integrity (collagen, keratin) to hold body tissue together, and biological catalysis to assist in cellular reactions, especially metabolism.

22.17 They both have the three-carbon skeleton found in Ala, but Phe has a phenyl group substituted on the β-methyl group:

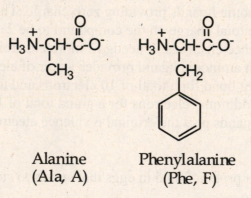

Alanine   Phenylalanine
(Ala, A)   (Phe, F)

22.19 Amino acids exist as zwitterions, compounds that have positive and negative charges on the same molecule. The molecules can interact with each other by strong electrostatic (ionic) bonds. The molecules can pack tightly together into crystalline forms like salts.

22.21 Amino acids have a carboxylic acid group that by its nature tends to donate a proton. They also contain a basic amino group that tends to accept a proton. These two groups will react in an acid/base reaction to form a zwitterion:

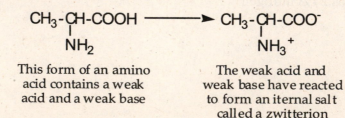

This form of an amino   The weak acid and
acid contains a weak   weak base have reacted
acid and a weak base   to form an iternal salt
                                called a zwitterion

22.23 The most predominant form at its isoelectric point will have the alpha amino and alpha carboxyl groups deprotonated and leave the side chain carboxyl protonated:

$$\overset{H}{\underset{\underset{\displaystyle COOH}{\overset{\displaystyle |}{CH_2}}}{\overset{\displaystyle |}{\underset{|}{C}}}} \quad H_3\overset{+}{N}-C-COO^-$$

$$H_3\overset{+}{N}-\underset{\underset{COOH}{\overset{|}{CH_2}}}{\overset{\overset{\displaystyle H}{|}}{C}}-COO^-$$

22.25 At its isoelectric point, lysine will have the alpha carboxyl group deprotonated, along with the alpha amino group. The side chain amino group will be protonated:

$$H_3\overset{+}{N}-\underset{\underset{-NH_3}{\overset{|}{(CH_2)_4}}}{\overset{\overset{\displaystyle H}{|}}{C}}-COO^-$$

22.27 The side chain imidazole is the most unique. All amino acids have an α-carboxyl, and all have an α-amino. Only histidine has a sidechain imidazole, a heterocyclic aromatic amine.

22.29 The side chain of histidine is an imidazole with a nitrogen that reversibly binds to a hydrogen. When dissociated it is neutral; when associated it is positive. Therefore, chemically it is a base, even though it does have a pKa in the acidic range.

22.31 Histidine, Arginine, and Lysine

22.33 Serine may be obtained by the hydroxylation of alanine. Tyrosine is obtained by the hydroxylation of phenylalanine.

22.35 Thyroxine is a hormone that stimulates overall metabolism.

22.37

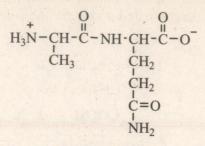

Alanylglutamine
(Ala-Gln)

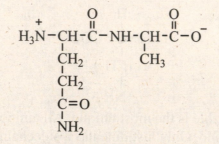

Glutaminylalanine
(Gln-Ala)

22.39 The tripeptide Thr-Arg-Met:

$$\underset{\underset{\text{CH}_3}{|}}{\underset{\underset{\text{CH-OH}}{|}}{\overset{+}{\text{H}_3\text{N}}-\text{CH}-\overset{\overset{\text{O}}{\|}}{\text{C}}}}-\text{NH}-\underset{\underset{\underset{\underset{\text{C}=\text{NH}_2^+}{|}}{\underset{\text{NH}}{|}}}{\underset{\underset{\text{CH}_2}{|}}{\underset{\text{CH}_2}{|}}}}{\underset{\text{CH}_2}{|}}\text{CH}-\overset{\overset{\text{O}}{\|}}{\text{C}}-\text{NH}-\underset{\underset{\underset{\text{CH}_3}{|}}{\underset{\text{S}}{|}}}{\underset{\underset{\text{CH}_2}{|}}{\underset{\text{CH}_2}{|}}}\text{CH}-\text{COO}^-$$

22.41 The side chains of the polypeptide of alternating Val and Phe are very nonpolar (only carbon and hydrogen atoms). The backbone, composed of the repeating pattern of peptide bonds, has atoms that show polarity. The N-H and O = C of the amide bonds may hydrogen bond with each other (as in an α-helix) or these groups may hydrogen bond with water.

22.43 (a) Structure of Met-Ser-Cys:

$$\overset{+}{H_3}N-CH-\overset{\overset{O}{\|}}{C}-NH-CH-\overset{\overset{O}{\|}}{C}-NH-CH-\overset{\overset{O}{\|}}{C}-O^-$$
$$\quad\;\; CH_2 \qquad\qquad CH_2OH \quad CH_2SH$$
$$\quad\;\; CH_2$$
$$\quad\;\; SCH_3$$

Met-Ser-Cys

(b) The pK_a of the carboxyl group is about 2, so at pH = 2:

$$\overset{+}{H_3}N-CH-\overset{\overset{O}{\|}}{C}-NH-CH-\overset{\overset{O}{\|}}{C}-NH-CH-\overset{\overset{O}{\|}}{C}-O^- \;+\; \overset{+}{H_3}N-CH-\overset{\overset{O}{\|}}{C}-NH-CH-\overset{\overset{O}{\|}}{C}-NH-CH-\overset{\overset{O}{\|}}{C}-OH$$
$$\quad\;\; CH_2 \qquad\qquad CH_2OH \quad CH_2SH \qquad\qquad\quad CH_2 \qquad\qquad CH_2OH \quad CH_2SH$$
$$\quad\;\; CH_2 \qquad\qquad\qquad\qquad\qquad\qquad\qquad\qquad\qquad CH_2$$
$$\quad\;\; SCH_3 \qquad\qquad\qquad\qquad\qquad\qquad\qquad\qquad\qquad SCH_3$$

The pK_a of the α-amino group is about 9:

$$\overset{+}{H_3}N-CH-\overset{\overset{O}{\|}}{C}-NH-CH-\overset{\overset{O}{\|}}{C}-NH-CH-\overset{\overset{O}{\|}}{C}-O^- \qquad \text{At pH 7.0 net charge = 0}$$
$$\quad\;\; CH_2 \qquad\qquad CH_2OH \quad CH_2SH$$
$$\quad\;\; CH_2$$
$$\quad\;\; SCH_3$$

$$H_2N-CH-\overset{\overset{O}{\|}}{C}-NH-CH-\overset{\overset{O}{\|}}{C}-NH-CH-\overset{\overset{O}{\|}}{C}-O^- \qquad \text{At pH 10, net charge = -1}$$
$$\quad\;\; CH_2 \qquad\qquad CH_2OH \quad CH_2SH$$
$$\quad\;\; CH_2$$
$$\quad\;\; SCH_3$$

22.45 The precipitated protein at its isoelectric point has an equal number of positive and negative charges (no net charge). The protein molecules clump together and form aggregates. Adding acid protonates the carboxylate groups on amino acid residue side chains and at the C- terminus giving the protein a net positive charge due to the protonated nitrogen atoms of the amino groups. The charged protein molecules now repel each other and will be more soluble. However, not all precipitations are reversible.

22.47 (a) The number of different tetrapeptides may be calculated by assuming that each of the four amino acids may be located in any of the four positions. The total number is 4 to the power of 4 or 256 possible tetrapeptides.

(b) If all 20 amino acids are available, the total number of possibilities is 20^4 or 160,000.

22.49 Leu, which has a side chain made up of only hydrocarbon, is a nonpolar amino acid. The fewest changes in the protein would result if the Leu were substituted by other nonpolar amino acids like Val or Ile.

22.51 (a) Tropocollagen, the triple helix form of collagen, is considered to be a form of secondary structure.

(b) The collagen fibril form is defined by quaternary structure.

(c) Collagen fibers are considered to be an example of quaternary structure.

(d) The repeating sequence of amino acids refers to the primary structure.

22.53 Iron is in group 8 of the periodic table, thus it has 8 valence electrons when it is uncharged. The compound $[Fe(CO)_5]$ is uncharged overall and each CO ligand is uncharged too. Thus, the iron is uncharged and needs 10 more electrons to achieve the 18 electron rule. These 10 valence electrons come from coordinate covalent bonds with the CO ligands.

22.55 $[Zn(NH_3)_2Cl_2]$ has no net charge and 2 chloride ligands each providing a 1-charge. The ammonia ligands are uncharged, so the charge on zinc must be 2+. Zinc is in group 12 of the periodic table so it has 12 valence electrons, but since the charge is 2+ in this case, it only has 10 valence electrons. This means that it needs 8 more electrons to satisfy the 18 electron rule. 4 electrons are provided by the 2 ammonia ligands and 4 more electrons are provided by the chloride ligands making a total of 18 electrons on the zinc atom.

22.57 The carboxylic acid side chain of Glu has a pK_a of 4.25. At low pH values, the side chain carboxyl groups are protonated and thus have no charge. At higher pH (above 4.25), the side chains begin to be deprotonated and become negatively charged. When the poly-peptide wraps into an α-helix, this brings the side chains in close proximity (Figure 22.10). Above a pH of 4.25, the negatively charged side chains interact unfavorably (repel each other) which destabilizes the helix, and the polypeptide forms a random coil. When the side chains are protonated at low pH, the neutral side chains allow for the formation of an α helix.

22.59 Box 1: the carboxyl terminus of the top polypeptide.
Box 2: the amino terminus of the lower polypeptide.
Box 3: a region of antiparallel β-pleated sheet between the top and bottom polypeptides.
Box 4: a section of random coil structure.
Box 5: a region of hydrophobic interactions.
Box 6: an intrachain disulfide bond between Cys residues.
Box 7: a region of α-helix.

Box 8: a salt bridge (ionic interaction) between the side chains of Asp and Lys.
Box 9: a hydrogen bond that is part of the tertiary structure.

22.61 (a) Adult hemoglobin (HbA) has a quaternary structure defined by four subunits, 2 α chains and 2 β chains. (The α and β terms are nomenclature only and do not refer to secondary structure in the protein). Fetal hemoglobin (HbF) also has four subunits, but instead of β subunits, HbF has γ subunits that have a different primary structure than the β subunits. The actual arrangement of the four subunits is very similar in HbA and HbF.
(b) HbF has a slightly greater affinity for oxygen than does HbA.
(c) Fetal hemoglobin has an oxygen saturation curve that is in between myoglobin and maternal hemoglobin, so the graph would look like the figure below:

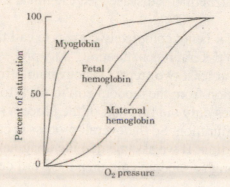

22.63 Cytochrome c is an example of a conjugated protein, one that contains, in addition to the protein portion, a non-amino acid portion called a prosthetic group. In cytochrome c the prosthetic group is an iron porphyrin, also called a heme. The two parts are held together by noncovalent interactions. Cytochrome c belongs to the conjugated protein group.

22.65 The urea disrupts hydrogen bonding between protein backbone C=O and H-N groups.

22.67 Cysteine residues in proteins are often linked via intramolecular or intermolecular disulfide bonds, -S-S-. When these bonds are selectively cleaved by reduction, the protein is usually denatured and the cysteine side chains become –SH.

22.69 Ions of heavy metals like silver denature bacterial proteins by reacting with cysteine -SH groups. The proteins, denatured by formation of silver salts, form insoluble precipitates.

22.71 Nutrasweet contains phenylalanine and many people are sensitive to this amino acid, especially when ingested in large quantities. It can cause severe headaches.

22.73 Hypoglycemic awareness is the sensation of hunger, sweating, and poor coordination that in a diabetic precedes an insulin reaction. This signals a time when insulin levels are too high relative to the blood sugar level (Chemical Connections 22C).

22.75 The α-helical regions of the prion protein are transformed into β-sheets, which tend to aggregate into amyloid plaques. These abnormal prions then stimulate other prions to alter their shape as well, which causes proliferation of the disease.

22.77 The binding curve of myoglobin is hyperbolic. The curve for hemoglobin is sigmoidal. At any partial pressure of oxygen, myoglobin is more highly saturated with oxygen than hemoglobin.

22.79 Diseases associated with amyloid plaques include mad cow disease, Creutzfeld-Jakob disease, and Alzheimer's disease (see Chemical Connections 22E).

22.81 Even if it is feasible, it is not completely correct to call the imaginary process that converts α-keratin to β-keratin denaturation. Any process that changes a protein from α to β requires at least two steps: (1) conversion from α form to a random coil, and (2) conversion from the random coil to the β form. The term denaturation describes only the first half of the process (Step 1). The second step would be called renaturation. So the overall process is called denaturation followed by renaturation. If we assume that the imaginary process actually occurs without passing through a random coil, then the word denaturation does not apply.

22.83 During aging, the triple helices of collagen are crosslinked by the formation of covalent bonds between lysine side chains. This is a form of quaternary structure; the subunits are crosslinked.

22.85 (a) Val and Ile, both with nonpolar side chains, interact by forming hydrophobic bonds.
(b) The side chains of Glu and Lys interact ionically to form a salt bridge.
(c) Hydrogen bonding between hydroxyl groups.
(d) Hydrophobic interactions between alkyl side chains

22.87 Gly has no stereocenter; therefore, it does not exist as enantiomers and does not rotate the plane of polarized light.

22.89 The amino acid Asp has three pK values: 1.88 for the α-carboxyl group; 4.25 for the side-chain carboxyl; 9.6 for the α-amino group. The forms present at pH 2:

$$HO-\overset{O}{\overset{\|}{C}}-CH_2-\underset{\underset{NH_3^+}{|}}{CH}-\overset{O}{\overset{\|}{C}}-O^-$$

$$HO-\overset{O}{\overset{\|}{C}}-CH_2-\underset{\underset{NH_3^+}{|}}{CH}-\overset{O}{\overset{\|}{C}}-OH$$

Major form present at
pH 2.0; net charge = 0

Minor form present at
pH 2.0; net charge = +1

22.91   The critical amino acids at the active site of an enzyme must be able to catalyze common organic chemistry reactions. Some side chains are more reactive than others. The acidic and basic amino acids are reactive. Serine has a hydroxyl in its side chain that is also very reactive.

22.93   Proteins can be denatured when the temperature is only slightly higher than a particular optimum. For this reason, the health of a warm-blooded animal is dependent on the body temperature. If the temperature is too high, proteins could denature and lose function.

22.95   If you know the genome of an organism, you do not necessarily know the proteins. Not all DNA encodes protein. Some DNA is never transcribed to RNA. Of the DNA that is transcribed to RNA, not all of the RNA is translated to protein. Even if a particular piece of DNA can be used to make RNA and then protein, the protein may not be made all of the time.

22.97   Such a diet supplement may help a person lose weight, but it would be of little use in repairing muscle tissue because collagen is a very incomplete protein. One third of its amino acids are glycine and another third are proline, both non-essential amino acids. Muscle repair requires high-quality protein rich in essential amino acids to be effective.

23.1   The word catalyst is a general term used to define agents that speed up chemical reactions. Enzymes are natural catalysts, composed of proteins and sometimes RNA, that catalyze reactions in biological cells.

23.3   Lipases, enzymes that catalyze the hydrolysis of ester bonds in triglycerides, are not very specific as they work on many different triglyceride structures with about the same effectiveness. It is predicted that the two triglycerides containing palmitic acid and oleic acid would be hydrolyzed at about the same rate.

23.5   There are thousands of different kinds of biochemicals in an organism and thousands of reactions are required to synthesize and degrade the compounds. The mild conditions inside the cell are not conducive to fast reaction rates so each reaction requires an enzyme for catalysis under physiological conditions.

23.7   Hydrolases use water to break bonds in a substrate usually leading to two smaller products; for example, the enzyme acetylcholinesterase in Table 23.1. Water is always involved in a hydrolase reaction. Lyases catalyze the addition of a group to a double bond or the removal of a group to form a double bond. The added or removed group may be water, but it is not always one of the substrates. An example of a lyase is aconitase in Table 23.1.

23.9   (a) The reactant and product have the same molecular formula, but they have different molecular structures, so they are isomers. The enzyme that catalyzes the reaction is classified as an isomerase.
(b) Water is a reactant that hydrolyzes the two amide bonds in urea. The enzyme is classified as a hydrolase.
(c) Succinate is oxidized and FAD is reduced. The enzyme is classified as an oxidoreductase.
(d) Aspartase catalyzes the addition of a functional group to a double bond. The enzyme is classified as a lyase.

23.11 A cofactor is a nonprotein part of an enzyme that must be present for catalytic activity. Metal ions are often cofactors. When a cofactor is an organic molecule like FAD or heme, it is called a coenzyme.

23.13 Noncompetitive inhibitors slow enzyme activity by binding to sites other than the active site where the substrate (S) binds. A reversible inhibitor (I) binds to and dissociates from the enzyme (E) thus setting up an equilibrium:

$$E + I \rightleftharpoons EI$$

An irreversible inhibitor binds permanently to the enzyme and alters its structure:

$$\text{E} + \text{I} \longrightarrow \text{E-I}$$

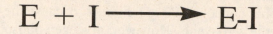

23.15 At very low concentrations of substrate, the rate of an enzyme-catalyzed reaction increases in a linear fashion with increasing substrate concentrations. At higher concentrations of substrate, the active sites of the enzyme molecules become saturated with substrate and the rate of increase is slowed. The rate is at a maximum level when all of the enzyme active sites are occupied with substrate. That maximum rate cannot be exceeded even if more substrate is added.

23.17 (a) Normal body temperature is 37° C (98.6° F) and a fever of 104° F is about 40° C. According to the curve, the bacterial enzyme is more active with the fever.
(b) The activity of the bacterial enzyme decreases if the patient's temperature decreases to 35° C.

23.19 (a) Plot of the pH dependence of pepsin activity:

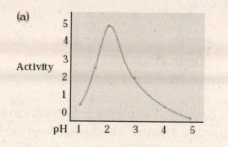

(b) Optimum pH is about 2
(c) Pepsin is predicted to have no catalytic activity at pH 7.4.

23.21 The normal substrate of urease, urea, is a much smaller molecule than diethylurea. Apparently the diethylurea, with its large bulky ethyl groups, is too large to fit into the active site of the enzyme

23.23 The amino acid residues most often found at enzyme active sites are His, Cys, Asp, Arg, and Glu.

23.25 The correct answer is (c). Initially the enzyme does not have just the right shape for strongly binding a substrate, but the shape of the active site changes to better accommodate the substrate molecule.

23.27 Amino acid residues in addition to those at an enzyme active site are present to help form a three-dimensional pocket where the substrate binds. These amino acids act to make the size, shape, and environment (polar or nonpolar) of the active site just right for the substrate.

23.29 At first, one might expect the inhibition of phosphorylase action by caffeine to be a case of traditional noncompetitive inhibition because the inhibitor apparently does not bind at the active site. However, glycogen phosphorylase is known to be an allosteric enzyme and the classical terms competitive and noncompetitive are not used to define inhibition. The term negative modulator is used to define the action of caffeine on the allosteric enzyme, phosphorylase.

23.31 The terms zymogen and proenzyme are both used to define an inactive form of an enzyme.

23.33 The action of a protein kinase is to catalyze the transfer of a phosphate group from ATP to another molecule, in this case a tyrosyl residue of an enzyme:

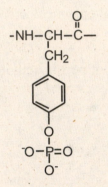

23.35 Glycogen phosphorylase b is converted to glycogen phosphorylase a by the action of an enzyme called phosphorylase kinase. The kinase transfers phosphate groups from ATP to phosphorylase b. Two ATPs are required because the phosphorylase has two subunits, each of which must be phosphorylated:

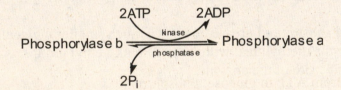

23.37 Glycogen Phosphorylase is controlled by allosteric regulation and by phosphorylation. The allosteric controls are very fast, so that when the level of ATP drops, for example, there is an immediate response to the enzyme allowing more energy to be produced. The covalent modifications by phosphorylation are triggered by hormone responses. They are a bit slower, but more long lasting and ultimately more effective.

23.39 Just like lactate dehydrogenase, there are 5 isozymes of PFK. They are designated as follows: M_4, M_3L, M_2L_2, ML_3, and L_4.

23.41 Two enzymes that increase in serum concentration following a heart attack are creatine phosphokinase and aspartate aminotransferase. Creatine phosphokinase peaks earlier than aspartate aminotransferase, so it would be the best choice in the first 24 hours.

23.43   Serum levels of the enzymes AST and ALT are monitored for diagnosis of hepatitis and heart attack. Serum levels of AST are increased after a heart attack, but ALT levels are normal. In hepatitis, both enzymes are elevated. The diagnosis, until further testing would indicate the patient may have had a heart attack.

23.45   Chemicals present in organic vapors are detoxified in the liver. The enzyme alkaline phosphatase is monitored to diagnose liver problems.

23.47   It is not possible to administer chymotrypsin orally. The stomach would treat it just as it does all dietary proteins: degrade it by hydrolysis to free amino acids. Even if whole, intact molecules of the enzyme were to be present in the stomach, the low pH in the region would not allow activity for the enzyme that has an optimum pH of 7.8.

23.49   A transition state analog is built to mimic the transition state of the reaction. It is not the same shape as the substrate or the product, but rather something in between. The potency with which such analogs can act as inhibitors lends credence to the theory of the induced fit.

23.51   Succinylcholine has a chemical structure similar to that of acetylcholine so they both can bind to the acetylcholine receptor of the muscle end plate. The binding of either choline causes a muscle contraction. However, the enzyme acetylcholinesterase hydrolyzes succinylcholine only very slowly compared to acetylcholine. Muscle contraction does not occur as long as succinylcholine is still present and acts as a relaxant.

23.53   Phosphofructokinase and hexokinase are two kinases found in glycolysis. In general, kinases act to transfer a phosphate from a high-energy phosphate, such as ATP, to another molecule, such as a sugar molecule. Hexokinase transfers a phosphate from ATP to glucose. Phosphfructokinase transfers a phosphate from ATP to fructose 6-phosphate.

23.55   Many people suffer mild to extreme effects from traumatic experiences in their lives. If scientists could selectively block these memories, they could alleviate much of the mental anguish of these people.

23.57   In the enzyme pyruvate kinase, the $=CH_2$ of the substrate phosphoenol pyruvate sits in a hydrophobic pocket formed by the amino acids Ala, Gly, and Thr. The methyl group on the side chain of Thr rather than the hydroxyl group is in the pocket. Hydrophobic interactions are at work here to hold the substrate into the active site.

23.59   Researchers were trying to inhibit phosphodiesterases because cGMP acts to relax constricted blood vessels. This was hoped to help treat angina and high blood pressure.

23.61   No, reducing the population of intestinal bacteria is not a good idea since they serve many important purposes. Researchers are instead working on an inhibitor for the bacterial enzyme glucoronidase which reacts with the colon cancer drug leading to the diarrhea side effect.

23.63   Phosphorylase exists in a phosphorylated form and an unphosphorylated form, with the former being more active. Phosphorylase is also controlled allosterically by several compounds, including AMP and glucose. While the two act semi-independently, they are related to some degree. The phosphorylated form has a higher tendency to assume the R form, which is more active, and the unphosphorylated form has a higher tendency to be in the less active T form.

23.65   Heme groups are coenzymes in several oxidoreductases including methylamine dehydrogenase (MADH) described in Chemical Connections 23F where they help shuttle electrons from one site to another.

23.67   In the processing of cocaine by specific esterase enzymes, the cocaine molecule passes through an intermediate state. A molecule was designed that mimics this transition state, and this transition state analog can be given to a host animal, which then produces antibodies to the analog. When these antibodies are given to a person, they act like an enzyme and degrade cocaine.

23.69   Cocaine blocks the reuptake of the neurotransmitter, dopamine, leading to overstimulation of the nervous system as the neurotransmitter will be around longer.

23.71   (a) Vegetables such as green beans, corn, and tomatoes are heated to kill microorganisms before they are preserved by canning. Milk is preserved by the heating process, pasteurization.
(b) Pickles and sauerkraut are preserved by storage in vinegar (acetic acid).

23.73   Active digestive enzymes should not be found in any cooked food including hot dogs, since the heat of cooking denatures and permanently inactivates most enzymes.

23.75   The amino acid residues (Lys and Arg) that are cleaved by trypsin have basic side chains, thus are positively charged at physiological pH.

23.77   This enzyme works best at a pH of about 7.

23.79   Chymotrypsin uses water as a substrate to cleave a peptide bond. Thus the reaction is hydrolysis and the enzyme is classified as a hydrolase.

23.81 (a) The reaction involves the oxidation of an alcohol, ethanol, to an aldehyde, acetaldehyde. The enzyme is called ethanol dehydrogenase or in general, alcohol dehydrogenase. It could also be called ethanol oxidoreductase.
(b) The reaction involves the hydrolysis of the ester bond in ethyl acetate. An appropriate name is ethyl acetate esterase or ethyl acetate hydrolase.

23.83 These enzymes, one from brain and one from liver, catalyze the same reaction, but exist in different forms. They would be called isoenzymes or isozymes.

23.85 No. All enzymes catalyze both the forward and reverse reactions in theory. In reality, however, the levels of substrates and products can often control the actual direction a reaction goes under physiological conditions. Enzymes cannot change the nature of thermodynamics. If there is a million to one ratio of substrate to product, for example, the enzyme cannot change the tendency of the substrates to form products.

23.87 These supplements in most cases have little value because they are taken orally and therefore have to pass through the digestive system where they are exposed to digestive enzymes which will degrade them. In addition, large molecules like proteins cannot pass through the intestinal wall. Even if these supplements were able to cross into the blood stream from the digestive system, there are no mechanisms for enzymes like superoxide dismutase to be taken up by cells.

23.89 A diuretic moves more water out of the body and into the bladder via its effect on the kidneys. This will tend to dehydrate a distance runner, so the beneficial effect on the lipases that help with energy production can be mitigated by the athlete being dehydrated. This effect will be greater in hot weather.

23.91 The structure of RNA makes it more likely to be able to adopt a wider range of tertiary structures, so it can fold up to form globular molecules like protein-based enzymes. It also has an extra oxygen, which gives it an additional reactive group to use in catalysis or an extra electronegative group useful in hydrogen bonding.

24.1   G-protein is a membrane-bound enzyme that catalyzes hydrolysis of GTP to GDP or GMP, whereas GTP is an energy storage molecule.

24.3   A chemical messenger is a biomolecule such as a hormone that binds to a receptor in a cell membrane with the purpose of initiating some change inside the cell. A secondary messenger is a biomolecule like cAMP that transmits the action of the chemical messenger to the inside of the cell, usually with amplification.

24.5   The role of calcium ions in the release of neurotransmitters is most easily understood when one considers a nerve impulse reaching the presynaptic site (Figure 24.4). This causes the opening of ion channels and the entry of calcium ions into the nerve cell that leads to the release of neurotransmitter molecules into the synapse.

24.7   Lactation is controlled by hormones from the anterior pituitary gland. The anterior lobe makes prolactin that stimulates the growth of the mammary gland. The posterior lobe of the pituitary produces the hormone oxytocin, which stimulates milk flow.

24.9   The neurotransmitter acetylcholine is stored in presynaptic vesicles (Fig 24.4). When a nerve impulse arrives, ion channels open and allow calcium ions to enter the nerve cell. The vesicles then fuse with the membrane and release acetylcholine into the synapse where the neurotransmitter molecules bind to receptors on the postsynaptic side. Binding of acetylcholine to the receptors opens ion channels. The flow of ions creates electrical signals.

24.11   Cobratoxin causes paralysis by acting as a nerve system antagonist—it blocks the receptor and interrupts the communication between neuron and muscle cell. The botulin toxin prevents the release of acetylcholine from the presynaptic vesicles.

24.13   Calcium sparks or puffs are sudden influxes of calcium ions from extracellular sources or intracellular stores, such as the endoplasmic reticulum. Calcium waves are slow fluctuations (increases and decreases) of calcium ion concentration inside the cell.

24.15   The calcium ion concentration must increase 100 to 250 times normal in order to achieve fusion.

24.17   Glutamic acid is removed from its receptor by a process called reuptake. A transporter protein molecule carries Glu back through the presynaptic membrane and into the neuron.

24.19 The structure of NMDA is shown in Section 24.4B. NMDA has a methyl group on its amino group. Asp does not have the methyl group. Also, the active enantiomer of NMDA is the D-form. Asp in proteins and organisms is the L-enantiomer.

24.21 The reaction for formation of cyclic AMP is shown in Section 24.5C. A phosphate anhydride bond between the α and β phosphorus atoms is broken and a new phosphate ester bond is formed between the α-phosphorus and the 3' hydroxyl group of ribose.

24.23 The conversion of GTP to cyclic GMP:

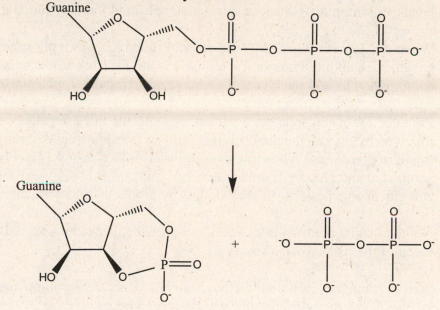

24.25 The binding of cyclic AMP activates the protein kinase by dissociating the regulatory and catalytic subunits (Figure 24.5).

24.27 The binding of acetylcholine molecules opens a channel for the flow of sodium and potassium ions. The movement of ions provides the electrical signal.

24.29 Histamine is removed from its receptors by the action of a monoamine oxidase. This results in the formation of an aldehyde that is taken up by the neuron and later converted back to histamine.

24.31 No, the opposite effects would be expected. Dramamine interacts with the H_1 type of histamine receptor, which is present in the respiratory tract and involved in asthma. Cimetidine interacts with the H_2 histamine receptors that are found in the stomach and involved in HCl release.

24.33   The drug Demerol relieves pain by acting as an agonist of brain enkephalins. The enkephalins are peptides that when bound to their receptors control pain perception. Demerol binds to the enkephalin receptors and acts in the same manner as these peptides.

24.35   Inositol-1,4,5-triphosphate is inactivated as a secondary messenger by hydrolytic removal of the phosphate group on position C-5. The reaction is catalyzed by a phosphatase (see Problem 24.34).

24.37   Glucagon is produced in the pancreas when this organ senses that blood glucose levels are dropping. Glucagon alters metabolism to increase blood glucose.

24.39   When cAMP activates cAMP-dependent Protein Kinase, it phosphorylates target enzymes, including Phosphofructokinase 2 and Fructose bisphosphatase 2. This combination lowers the levels of the molecule, fructose 2,6 bisphosphate.

24.41   Fructose 2,6 bisphosphate is a regulatory molecule that affects key enzymes involved in glucose metabolism. It stimulates phosphofructokinase, an enzyme found in glycolysis. When fructose 2,6 bisphosphate is high, glycolysis is stimulated and glucose is used for energy. When it is low, glycolysis is inhibited and gluconeogenesis is stimulated, thereby producing more glucose.

24.43   No, the insulin pathway does not use G-proteins. Instead it uses a different type of receptor that is a tyrosine kinase.

24.45   Steroid hormones can bind to DNA and thus affect transcription and protein synthesis, but the response time can be several hours.

24.47   The first theory is that neurons are active during sleep reinforcing synaptic connections made during the day. The second theory is that neurons rest during sleep and are inactive.

24.49   The botulin toxin, found in *C. botulinum*, prevents the release of acetylcholine from the presynaptic vesicle. No neurotransmitter reaches the receptor molecules.

24.51   The neurofibrillar tangles found in the brains of Alzheimer's patients are composed of tau proteins. Mutated tau proteins, which normally interact with the cytoskeleton, grow into these tangles instead, thus altering normal cell structure.

24.53 Drugs that increase the concentration of the neurotransmitter acetylcholine may be effective in the treatment of Alzheimer's disease. Acetylcholinesterase inhibitors, such as Aricept, inhibit the enzyme that decomposes the neurotransmitter.

24.55 There is a group of 25 extended families with more than 5,000 total members. These families develop Alzheimer's before the age of 50 which is 15 to 20 years earlier than the average case. This makes it much easier to test preventative and therapeutic therapies since there is a very large known pool of susceptible individuals.

24.57 First, 5 to 20 years before symptoms are apparent, β-amyloid aggregates form in the areas of the brain involved in making new memories. This hinders nerve transmission in the brain by blocking neurotransmitter receptors. Then 1 to 5 years before symptoms, tau proteins begin to build up and become phosphorylated. Phosphorylated tau proteins lead to microtubule destruction and formation of toxic tau protein tangles. Lastly, 1 to 5 years before diagnosis, the brain actually begin to shrink, especially areas involved in new memories like the hippocampus.

24.59 The neurotransmitter dopamine is deficient in patients of Parkinson's disease. The biochemical cannot be administered orally, because the molecules cannot cross the blood-brain barrier.

24.61 Drugs like Cogentin that block cholinergic receptors are often used to treat the symptoms of Parkinson's disease. These drugs lessen spastic motions and tremors.

24.63 Nitric oxide relaxes the smooth muscle cells that surround blood vessels. This causes increased blood flow in the brain that causes headaches.

24.65 A blocked artery caused by a stroke restricts the blood flow to certain parts of the brain which kills neurons. Adjoining neurons begin to release glutamic acid and NO which kills local cells.

24.67 Insulin-dependent diabetes is caused by insufficient production of insulin by the pancreas. The administration of insulin relieves symptoms of this type of diabetes. Noninsulin-dependent diabetes is caused by a deficiency of insulin receptors or by the presence of inactive receptor molecules. These patients do not respond to injected insulin so they use certain drugs to relieve symptoms.

24.69   The new technique of monitoring glucose levels in the tear relieves the patient of pricking his/her finger many times a day for a blood sample.

24.71   Moderate physical training greatly increases the level of GLUT4 receptors in muscle cells. Studies showed that 1 week of moderate exercising doubled the muscle cell GLUT4 content of sedentary individuals.

24.73   When scientists observed moderately fit middle-aged men who stopped exercising as part of the study, they observed that it only took a few days for GLUT4 levels to drop to less than half their normal value.

24.75   Bisphenol A has been linked to enlargement of the prostate and increased chromosomal abnormalities in mice. It has also been linked to breast cancer and early onset of puberty in humans.

24.77   Effects of NO on smooth muscle: dilation of blood vessels and increased blood flow; headaches caused by dilation of blood vessels in brain; increased blood flow in the penis leading to erections.

24.79   Aldosterone is a steroid hormone that acts by binding to steroid receptors in the nucleus. This complex of steroid and receptor becomes a transcription factor that regulates expression of a gene that codes proteins for mineral metabolism.

24.81   Administering large doses of acetylcholine will ameliorate the overdose of decamethonium bromide. In competitive inhibition the inhibitor can be removed by increase in the substrate concentration. The esterase may regain its activity.

24.83   See structures for α-alanine and β-alanine below. In the normal α-alanine, both the carboxyl and amino groups are bound to the same carbon atom. In β-alanine, the two functional groups are bound to different carbon atoms.

$$CH_3-\overset{\alpha}{\underset{\underset{NH_3^+}{|}}{C}H}-COO^-$$ $$\overset{\beta}{\underset{\underset{NH_3^+}{|}}{C}H_2}-CH_2-COO^-$$

Alanine   β-Alanine
(an α-amino acid)   (a β-amino acid)

24.85   (a) The hormone vasopressin and the neurotransmitter dopamine have little in common except that they both elicit their actions by binding to specific receptor proteins.
(b) Vasopressin binds to a receptor that activates protein kinase C and involves the second messenger phosphatidyl inositol. Dopamine is a neurotransmitter that is amplified by cyclic AMP secondary messenger.

24.87   Cholera toxin permanently activates the G-protein which leads to overproduction of cyclic AMP and continuously opens ion channels. This results in the constant outflow of ions and accompanying water causing dehydration and diarrhea.

24.89   Glucose levels in the serum would decrease as the sugar is rapidly taken up by cells.

24.91   Proteins are capable of specific interactions at recognition sites, as we saw in chapter 23. This makes proteins ideal for the selectivity receptors must show for particular messengers.

24.93   Adrenergic messengers, such as dopamine, are derivatives of amino acids. For example, a biochemical pathway exists that produces dopamine from the amino acid tyrosine.

24.95   Insulin is a small protein. It would go through protein digestion if taken orally and would not be taken up as the whole protein. In general, peptide hormones are ineffective when administered orally.

24.97   Steroid hormones can directly affect nucleic acid synthesis.

24.99   Chemical messengers vary in their response times. Those that operate over short distances, such as neurotransmitters, have short response times. Their mode of action frequently consists of opening or closing channels in a membrane or binding to a membrane-bound receptor. Hormones must be transmitted in the blood stream, which requires a longer time for them to take effect. Some hormones can and do affect protein synthesis, which makes the response time even longer.

24.101 Having two different enzymes for the synthesis and breakdown of acetylcholine means that the rates of formation and breakdown can be controlled independently.

25.1   Structure of UMP:

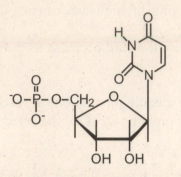

25.3   Scientists and physicians have identified hundreds of hereditary diseases. Sickle cell anemia is one you have studied in detail (Chemical Connections 22D).

25.5   (a) In eukaryotic cells, DNA is located in the cell nucleus and in mitochondria. (b) RNA is synthesized from DNA in the nucleus, but further use of RNA (protein synthesis) occurs on ribosomes in the cytoplasm.

25.7   DNA has the sugar deoxyribose, while RNA has the sugar ribose. Also, RNA has uracil, while DNA has thymine.

25.9   See Figure 25.1. Thymine and uracil are both based on the pyrimidine ring. However, thymine has a methyl substituent at carbon 5 whereas uracil has a H. All of the other ring substituents are the same.

25.11   (a) Structure of cytidine: (b) Structure of deoxycytidine:

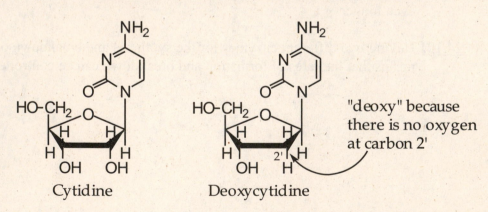

Cytidine   Deoxycytidine

"deoxy" because there is no oxygen at carbon 2'

25.13   D-ribose and 2-deoxy-D-ribose have the same structure except at carbon 2. D-ribose has a hydroxyl group and hydrogen on carbon 2, whereas deoxyribose has two hydrogens (Section 25.2B):

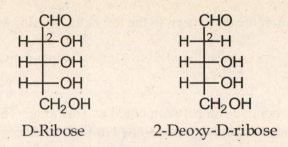

D-Ribose   2-Deoxy-D-ribose

25.15 The name nucleic acid comes from the fact that the nucleosides are linked by phosphate groups, which are the dissociated form of phosphoric acid.

25.17 The bond between the two phosphate groups is called an anhydride bond.

25.19 In RNA, carbons 3' and 5' of the ribose are linked by ester bonds to the phosphates. Carbon 1 is linked to the nitrogen base with the N-glycosidic bond.

25.21 (a) Structure of UDP:

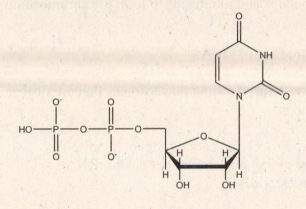

(b) Structure of dAMP:

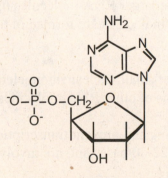

25.23 (a) One end will have a free 5' phosphate or hydroxyl group that is not in phospho-diester linkage. That end is called the 5' end. The other end, the 3' end, will have a 3' free phosphate or hydroxyl group.

(b) By convention the end drawn to the left is the 5' end. A is the 5' end and C is the 3' end.

(c) Using the same convention of 5' to 3', the complementary sequence is GTATTGCCAT.

25.25 Two hydrogen bonds form between uracil and adenine. This base pairing is the same as with adenine and thymine shown in Figure 25.5.

25.27 Histones are proteins with a high content of two basic amino acids, Lys and Arg. Recall that these two amino acid residues in proteins will have positively-charged side chains at physiological pH. On the other hand, DNA at physiological pH will have many negatively charged groups due to the ionized phosphates in the backbone. These two types of molecules will form very strong electrostatic interactions or salt bridges.

25.29 The superstructure of chromosomes is comprised of many elements. DNA and histones combine to form nucleosomes that are wound into chromatin fibers. These fibers are further twisted into loops and minibands to form the chromosome superstructure. (Figure 25.8).

25.31 The double helix is the secondary structure of DNA.

25.33 DNA is wound around histones, collectively forming nucleosomes that are further wound into solenoids, loops, and bands.

25.35 tRNA molecules range in size from 73 to 93 nucleotides per chain. mRNA molecules are larger with an average of 750 nucleotides per chain. rRNA molecules can be as large as 3000 nucleotides.

25.37 All types of RNA will have a sequence complementary to a portion of a DNA molecule because the RNA is transcribed from a gene.

25.39 Ribozymes, or catalytic forms of RNA, are involved in post-transcriptional splicing reactions that cleave larger RNA molecules into smaller more active forms. For example, tRNA molecules are formed in this way. Ribozymes are also part of protein synthesis.

25.41 Small nuclear RNA is part of small nuclear ribonucleoprotein particles which are involved in splicing reactions of other RNA molecules.

25.43 micro RNAs are 22 bases long and prevent transcription of certain genes. Small interfering RNA vary from 22-30 bases and are involved in the degradation of specific mRNA molecules.

25.45   Messenger RNA immediately after transcription contains both introns and exons. The introns are cleaved out by the action of ribozymes that catalyze splicing reactions on the mRNA.

25.47   Satellite DNA, in which short nucleotide sequences are repeated hundreds and even thousands of times, are not expressed into proteins. They are found at the ends and centers of chromosomes and are necessary for stability.

25.49   The greatest single safeguard against errors in DNA duplication is the high level of specificity between the base pairs, A-T and G-C. This protects the replication process from bringing in the wrong bases. This specificity shows strong molecular recognition.

25.51

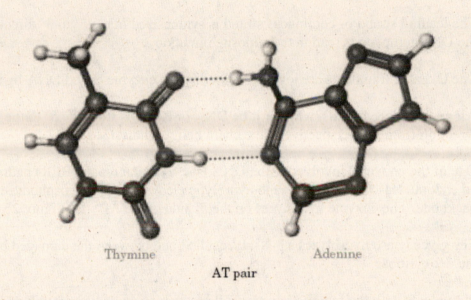

Thymine   Adenine

AT pair

25.53   A DNA double helix has four different kinds of bases: A, T, C, and G.

25.55   In semiconservative DNA replication, the original DNA (parental strand) separates and each strand is then used as the template to make its complementary strand. The product is two daughter molecules, each of which have one parental strand and one new strand. This is the mechanism of DNA replication in cells.

25.57   Before the process of DNA duplication can actually begin, the superstructure of the chromosome must be opened so the DNA molecule is accessible for copying. Part of this unwinding process is carried out by histone deacetylase that removes acetyl groups from lysine residue side chains of the histones. The reaction is shown in Section 25.6.

$$\text{Histone}-(CH_2)_4-NH_3{}^+ + CH_3-COO^- \underset{\text{deacetylation}}{\overset{\text{acetylation}}{\rightleftharpoons}}$$

$$\text{Histone}-(CH_2)_4-NH-\overset{\displaystyle O}{\overset{\|}{C}}-CH_3$$

25.59   Helicases are enzymes that break the hydrogen bonding between the base pairs in double helix DNA and thus help the helix to unwind. This prepares the DNA for the replication process.

25.61   The primers for starting DNA synthesis are short strands of RNA made from ribonucleoside triphosphates. Pyrophosphate (P-P) is a side product formed during the synthesis of the primers.

25.63   The leading strand or continuous strand is synthesized in the 5' to 3' direction so the DNA template is read in the opposite direction.

25.65   The Okazaki fragments are joined together by the enzyme called DNA ligase.

25.67   All nucleotide synthesis, whether it be DNA or RNA, is from the 5' to the 3' direction from the perspective of the chain being synthesized.

25.69   One of the enzymes involved in the DNA base excision repair (BER) pathway is an endonuclease that catalyzes the hydrolytic cleavage of the phosphodiester backbone. The enzyme hydrolyzes on the 5' side of the AP site (Figure 25.15).

25.71   The glycosylase hydrolyzes a β-N-glycosidic bond between the damaged base and the deoxyribose.

25.73   Individuals with the inherited disease XP are lacking an enzyme involved in the NER pathway. They are not able to make repairs in DNA damaged by UV light.

25.75   The 12-nucleotide primer to use is: 5'ATGGCAGTAGGC3'.

25.77   The first landmark case on the subject of biological patenting took place in the 1970s. A G.E. engineer attempted to patent a bacterial strain that was capable of breaking down oil slicks efficiently. Initially, the patent office rejected the application on the grounds that life forms and natural products could not be patented. The case was appealed all the way to the Supreme Court, which reversed the patent office decision by stating that one could patent "anything under the sun that is made by man."

25.79   Biotech firms own the purified DNA sequence and the techniques to study the gene, not the actual DNA in your body.

25.81   DNA polymerase, the enzyme that makes the phosphodiester bonds in DNA, is not able to work at the end of linear DNA. This results in the shortening of the telomeres at each replication. The telomere shortening acts as a timer for the cell to keep track of the number of divisions.

25.83   Because the genome is circular, even if the 5' primers are removed, there will always be DNA upstream that can act as a primer for DNA polymerase to use to synthesize DNA.

25.85   DNA samples from mother, child, and potential father are amplified by PCR, cleaved into fragments using restriction enzymes, and the fragments separated and analyzed by electrophoresis. The figure in CC 25C shows the results of these procedures. Lanes 2, 3, and 4 show the results for the father, child, and mother respectively. The child's DNA contains 6 bands and the mother's has five bands, all of which match those of the child, confirming the mother:child relationship. The band from the alleged father also contains six bands of which only three match the DNA of the child. A child is expected to inherit half of his/her genes from the father. These results do not exclude the potential father because every band in the child's DNA shows up in either the mother lane or the father lane.

25.87   Once a DNA fingerprint is made, each band in the child must come from one of the parents. Therefore, if the child has a band, and the mother does not, then the father must have that band. In this way, possible fathers are eliminated.

25.89   People fear that if your genotype fell into the wrong hands, it could be used against you. For example, an insurance company could refuse to insure you, or could give you much higher rates based on your genotype. Employers could refuse to hire you if you had a predisposition for a disease or other ailment that might make you a less efficient worker. Ultimately, a class system based on genotype could develop.

25.91   Venter's experiments on synthetic life proceeded in 3 phases. First, in 2008 Venter and colleagues demonstrated that they could synthesize an artificial chromosome of the bacterium *Mycoplasma genitalium* that included all 600,000 genes plus some "watermarks" – artificial sequences that allowed them to differentiate between the natural chromosome and the artificial one. Then in 2009, they demonstrated that they could move the genome purified from one bacterial species, *Mycoplasma mycoides* into another closely related species, *M. capricolum* effectively converting that into *M. mycoides*. In the final phase in 2010, they synthesized the *M. mycoides* genome and transplanted it into a *M. capricolum* that had its entire genome removed, once again converting the *M. capricolum* into *M. mycoides*. The watermark sequences in the synthesized *M. mycoides* DNA caused the bacteria to turn blue demonstrating that the transplanted DNA had successfully replaced the original DNA.

25.93   The argument is that since modern-day humans of European and Asian ancestry have between 1 – 4% Neandertal DNA in them, Neandertals are not truly extinct.

25.95   NRG3 involved in schizophrenia and CADPS2, AUTS2 both involved in autism are different in Neandertals.

25.97   The active site of a ribozyme is a three-dimensional pocket of ribonucleotides where substrate molecules are bound for catalytic reaction. Functional groups for catalysis include the phosphate backbone, ribose hydroxyl groups, and the nitrogen bases.

25.99   (a) The structure of the nitrogen base uracil is shown in Figure 25.1. It is a component of RNA.
(b) Uracil with a ribose attached by an N-glycosidic bond is called uridine.

25.101 Native DNA is the largest nucleic acid.

25.103 Mol % A = 29.3; Mol % T = 29.3; Mol % G = 20.7; Mol % C = 20.7.

25.105 RNA synthesis, like DNA synthesis, is 5' to 3' from the perspective of the growing chain. This is because the nature of the polymerase reaction is the nucleophilic attack of the 3' hydroxyl group on the 5' phosphate of another nucleotide.

25.107 DNA replication requires a primer, which is RNA. Since RNA synthesis does not require a primer, it makes sense that RNA must have preceded DNA as a genetic material. This added to the fact that RNA has been shown to be able to catalyze reactions means that RNA can be both an enzyme and a heredity molecule.

25.109 Guanine and cytosine make 3 hydrogen bonds between each other when the bond in DNA. Adenine and thymine make only 2 hydrogen bonds. Therefore, it takes more energy to break a G-C bond than it does an A-T bond, so the energy necessary to separate strands of DNA is proportional to the number of G-C bonds.

25.111 DNA synthesis must have extensive proofreading in order to make cells that are identical during replication. If errors are not detected, the mistake can be passed on from generation to generation. With RNA synthesis, if a mistake is made, only one product molecule is affected, but the next time the DNA is transcribed to RNA, it will make a correct one. An analogy would be a cook book. If the cook book has a recipe that is printed wrong, then every time the recipe is followed, the dish created will be wrong. On the other hand if the book is correct but you read it wrong one time, you will make the dish incorrectly one time but will probably do it correctly the next time.

Chapter 26: Gene Expression and Protein Synthesis

26.1   Transcription begins when the DNA double helix begins to unwind at a point near the gene that is to be transcribed. The superstructures of DNA (chromosomes, chromatin, etc.) break down to the nucleosome (histones plus DNA). Binding proteins interact with the nucleosomes making the DNA less dense and more accessible. The helicase enzymes then begin to unwind the double helix.

26.3   Valine + tRNA (specific for Val) + ATP

26.5   Answer (c); gene expression refers to both processes, transcription and translation.

26.7   Protein translation occurs on the ribosomes.

26.9   Helicases are enzymes that catalyze the unwinding of the DNA double helix prior to transcription. The helicases break the hydrogen bonds between base pairs.

26.11   The termination signal for transcription is at the 5' end of the DNA.

26.13   The "guanine cap" methyl group is located on nitrogen number 7 of guanine.

26.15   DNA replication requires an RNA primer, however RNA transcription does not require a primer.

26.17   Consensus sequences are non-unique nucleotide sequences like the TATA box that are found within the promoter 26 base pairs upstream from the transcriptional start site.

26.19   An intron is a portion of the structural gene that is removed post-transcriptionally by splicing (see Section 26.3). It does not encode protein

26.21   A codon, the three-nucleotide sequence that specifies amino acids for protein synthesis, is located on a mRNA molecule.

26.23   The main subunits are the 60S and the 40S ribosomal subunits, although these can be dissociated into even smaller subunits.

26.25   Each triplet sequence of the DNA has the code for a single amino acid in the protein. Therefore, the total number of amino acids is $981 \div 3 = 327$.

26.27   Leucine, arginine, and serine have the most, with 6 codons apiece. Methionine and tryptophan have the fewest with one apiece.

26.29   A genetic code that only had 2 nucleotides per codon could only encode 4^2 or 16 amino acids (where the number 4 is the number of different nucleotides and the exponent 2 is the length of the codon). This would be insufficient since there are 20 different amino acids.

26.31   The amino acid for protein translation is linked via an ester bond to the 3' end of the tRNA. The energy for producing the ester bond comes from breaking two, energy-rich phosphate anhydride bonds in ATP (producing AMP and two phosphates).

26.33   (a) The 40S subunit in eukaryotes forms the pre-initiation complex with the mRNA and the Met-tRNA that will become the first amino acid in the protein. (b) The 60S subunit binds to the pre-initiation complex and brings in the next aminoacyl-tRNA. The 60S subunit contains the peptidyl transferase enzyme.

26.35   Elongation factors are proteins that participate in the process of tRNA binding and movement of the ribosome on the mRNA during the elongation process in translation.

26.37   There is a special tRNA molecule used for initiating protein synthesis. In prokaryotes, this is $tRNA^{fmet}$, which will carry a formyl-methionine. In eukaryotes, there is a similar one, but it carries methionine. However, this tRNA carrying methionine for the initiation of synthesis is different from the tRNA carrying methionine for internal positions.

26.39   This is because it has been shown that there are no amino acids in the vicinity of the nucleophilic attack that leads to peptide bond formation. Therefore, the ribosome must be using its RNA portion to catalyze the reaction, so it is a type of enzyme called a ribozyme.

26.41   The parts of DNA involved in control of transcription are promoters, enhancers, silencers, and response elements. Molecules that bind to the DNA are RNA polymerase and a variety of transcription factors.

26.43   The active site of aminoacyl-tRNA synthases (AARS) contains the sieving portions that act to make sure that each amino acid is linked to its correct tRNA. There are two sieving steps that work on the basis of the size of the amino acid.

26.45   Both are DNA sequences that bind to transcription factors. The difference is largely due to our own understanding of the big picture. A response element is understood to control a set of responses in a particular metabolic context, such as a response element that activates several genes when the organism is challenged metabolically by heavy metals, by heat, or by a reduction in oxygen pressure.

26.47   Proteosomes are cylindrical assemblies of a number of protein subunits with proteolytic activity. Proteosomes play a role in post-translational degradation of damaged proteins. Proteins damaged by age or proteins that have misfolded are degraded by the proteosomes.

26.49 (a) Silent mutation: assume the DNA sequence is TAT on the coding strand, which will lead to UAU on the mRNA. Tyrosine is incorporated into the protein. Now assume a mutation in the DNA to TAC. This will lead to UAC in mRNA. Again, the amino acid will be tyrosine.
(b) Lethal mutation: the original DNA sequence on the coding strand is GAA, which will lead to GAA on mRNA. This codes for the amino acid glutamic acid. The DNA mutation TAA will lead to UAA, a stop signal that incorporates no amino acid. This could mean a vital protein is not made and could lead to disease or even a non-viable organism.

26.51 Yes, a harmful mutation may be carried as a recessive gene from generation to generation with no individual demonstrating symptoms of the disease. Only when both parents carry recessive genes does an offspring have a 25% chance of inheriting the disease.

26.53 Restriction endonucleases are enzymes that recognize specific sequences on DNA and catalyze the hydrolysis of phosphodiester bonds in that region thus cleaving both strands of the DNA (see Section 26.8). These enzymes are useful for the preparation of recombinant DNA.

26.55 Mutation by natural selection is an exceedingly long, slow process that has occurred for centuries. Each natural change in the gene has been ecologically tested and found usually to have a positive effect or the organism is not viable. Genetic engineering, where a DNA mutation is done very fast, does not provide sufficient time to observe all of the possible biological and ecological consequences of the change.

26.57 Modern molecular biology is based heavily on the ability to cut pieces of nucleic acids out and put them into other pieces of DNA, such as cutting out a gene of interest and splicing it into a plasmid or virus. This allows the DNA to be amplified and studied. Until the discovery of restriction enzymes, scientists could not cut chunks of DNA in a specific way at a specific place. They also could not put those pieces of DNA into a vector for amplification or transfer.

26.59 The viral coat is a protective protein covering around a virus particle.

26.61 An invariant site is a location in a protein that has the same amino acid in all species that have been studied. Studies of invariant sites help establish genetic links and evolutionary relationships.

26.63 The codon UCU specifies serine. In fact, all codons that begin with UC specify serine, so the old dogma would have been that any change in the third base would be a silent mutation as the amino acid would always be serine. However, we now know that it is possible to have a non-silent mutation with this situation. Although any third base will lead to serine, it may still require a different tRNA to be incorporated during translation. Ribosomes do not translate all versions of a codon at the same speed, which can affect the protein product.

26.65 The speed with which a ribosome translates the mRNA can be dependent on which of the codon possibilities for a particular amino acid are encountered. If the ribosome goes more quickly or more slowly, sometimes the protein product will fold incorrectly, leading to a protein that does not function the same.

26.67 The protein p53 is a tumor suppressor. When the protein's gene is mutated, the protein no longer controls replication and the cell begins to grow at an increased rate.

26.69 The Duffy protein is a protein found on the red blood cells and which acts as a docking site for malaria carrying parasites. Lack of the Duffy protein is beneficial to individuals living in malaria-prone regions.

26.71 Gene X could have a mutation in it such that the eventual protein product was altered, such as changing one amino acid for another. Gene X could have a silent mutation in it that nonetheless led to an altered protein due to altered translation of the gene. There could be a mutation in a region of the DNA upstream of Gene X that affected the ability of RNA polymerase to bind. There could be a mutation in a section of DNA that binds an enhancer or a silencer.

26.73 Since cystic fibrosis is caused by a genetic mutation that affects the lungs, it was thought that an adenovirus gene therapy vector (which readily infects lung tissue) carrying the correct gene could be used to infect the lungs of CF patients and cure them.

26.75 By isolating the CFTR gene, scientists could find the specific mutations that cause disease. They also could use the gene to help develop therapies to treat the disease.

26.77 VX-809 helps the mutant CFTR protein get to the cell surface and VX-770 improves the function of the mutant CFTR. Both drugs help the misfolded CFTR function more effectively, increasing lung function.

26.79 (a) Transcription: the units include the DNA being transcribed, the RNA polymerases, and a variety of general transcription factors.
(b) Translation: mRNA, ribosomal subunits, aminoacyl-tRNA, initiation factors, and elongation factors.

26.81 Hereditary diseases cannot be prevented, but genetic counseling can help people understand the risks involved in passing a mutated gene to offspring.

26.83   (a) Plasmid: a small, closed circular piece of DNA found in bacteria. It is replicated in a process independent of the bacterial chromosome.
(b) Gene: a section of chromosomal DNA that codes for a particular protein molecule or RNA.

26.85   Each of the amino acids has four codons. All the codons start with G. The second base is different for each amino acid. The third base may be any of the four possible bases. The distinguishing feature for each amino acid is the second base.

26.87   The hexapeptide is Ala-Glu-Val-Glu-Val-Trp.

27.1   Chemical energy present in food molecules is extracted and converted to a usable form by the process of catabolism. The energy derived from the degradation of different types of molecules is collected in the form of the energy-rich molecule, ATP.

27.3   (a) Mitochondria have two membranes, a highly-folded inner membrane, and an outer membrane (see Figure 27.3).
(b) The outer membrane is permeable to the diffusion of small ions and molecules. Special transport processes are required to move molecules through the inner membrane.

27.5   Cristae are the folds that are present in the inner mitochondrial membrane. The folds provide extensive surface area for the concentration of enzymes and other components required for metabolism.

27.7   Each ATP molecule has two phosphate anhydride bonds that release a substantial amount of energy when they are hydrolyzed during metabolic processes (Figure 27.5):

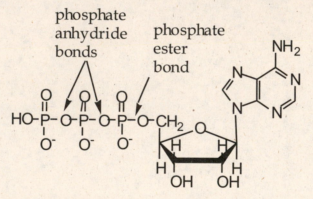

27.9   In reactions (a) and (b) the same type of anhydride bond is hydrolyzed. When the reactions are measured under standard conditions, the energy yield is about the same, 7.3 kcal/mole. In the cell, however, the concentration of ADP is very low, compared to ATP, and ADP is rarely used for energy.

27.11   The chemical bond between the ribitol and flavin in FAD is an amine (see Figure 27.6).

27.13   Two nitrogen atoms in the flavin ring that are linked to carbon (N=C) are reduced to yield $FADH_2$.

27.15   (a) The most important carrier of phosphate groups is ATP.
(b) The most important carriers of hydrogen ions and electrons from redox reactions are NADH and $FADH_2$.
(c) Coenzyme A carries acetyl groups.

27.17   An amide linkage is formed to bring together the amine group of mercaptoethylamine and the carboxyl group of pantothenic acid (Figure 27.7).

27.19   No, the reactive part of CoA is the thiol group (-SH) at the end of the molecule.

27.21   Most fats and carbohydrates are degraded in catabolism to the compound acetyl CoA.

27.23   α-Ketoglutarate is the only C-5 intermediate in the citric acid cycle.

27.25   FAD is used in the citric acid cycle as a coenzyme to oxidize succinate to fumarate.

27.27   Fumarase is classified as a lyase because it catalyzes the addition of water to a double bond.

27.29   ATP is not produced directly by the citric acid cycle. GTP, a reactive molecule with the same amount of energy as ATP, is produced in Step 5 and may be used for some energy- requiring processes.

27.31   The stepwise degradation and oxidation of acetyl CoA in the citric acid cycle is very efficient in the extraction and collection of energy. Rather than occurring in one single burst, the energy is released in small increments and collected in the form of reduced cofactors, NADH and $FADH_2$.

27.33   Carbon-carbon double bonds are present in the citric acid cycle intermediates fumarate and *cis*-aconitate.

27.35   When α-ketoglutarate is oxidized in Step 4 of the citric acid cycle, the electrons are transferred to NAD^+ to make the reduced form, NADH.

27.37   The mobile carriers of electrons in the electron transport chain are coenzyme Q and cytochrome *c*.

27.39   The proton translocator ATPase is a complex that resembles a rotor engine. It rotates every time a proton passes through the inner membrane (Figure 27.10).

27.41   During oxidative phosphorylation water is formed from protons, electrons, and oxygen on the matrix side of the inner membrane. This occurs when electrons are shuttled through complex IV (Figure 27.10).

27.43   (a) For each pair of protons translocated through the ATPase complex, one molecule of ATP is generated. Each pair of electrons that enters oxidative phosphorylation at complex I yields three ATP.
    (b) Each C-2 fragment, which represents the carbons in acetyl CoA, yields 12 ATP.

27.45 Protons are translocated through a proton channel formed by the Fo part of the ATPase which has 12 protein subunits embedded in the inner membrane (Figure 26.10).

27.47 The catalytic unit of ATPase is composed of α and β subunits (Figure 27.10). This part of the ATPase catalyzes the formation of ATP:

$$ADP + P_i \longrightarrow ATP + H_2O$$

27.49 The energy, in kcal, from the oxidation of 1 g of acetate by the citric acid cycle: Molecular wt. acetate = 59 g/mole. 1 g of acetate = 1 ÷ 59 = 0.017 mole. Each mole of acetate produces 12 moles of ATP (See Problem 27.43(b)).
0.017 mole × 12 = 0.204 mole of ATP.
0.204 mole of ATP × 7.3 kcal/mole = 1.5 kcal.

27.51 (a) Muscle contraction takes place when thick protein filaments called myosin slide past thin protein filaments called actin. The hydrolysis of ATP by myosin, an ATPase, drives the alternating association and dissociation of actin and myosin. This causes the contraction and relaxation of muscles (Section 27.8C). (b) The energy in muscle contraction comes from ATP.

27.53 Glycogen phosphorylase is activated by phosphorylation of serine residues in the protein subunits. The phosphoryl groups are transferred from ATP (see Chemical Connection 23E). Phosphorylase is also activated by allosteric effectors.

27.55 The antibiotic oligomycin inhibits the catalytic subunits of ATPase, thus it stops phosphorylation of ADP. Although it acts as an effective antibiotic, its toxic effect on ATPase does not allow its use in humans.

27.57 Number of g of acetic acid that must be metabolized to yield 87.6 kcal of energy: 87.6 kcal ÷ 7.3 kcal/mole ATP = 12 moles of ATP. 12 moles of ATP are released from the oxidation of 1 mole of acetate. 1 mole of acetate (MW = 60) = 60 g of acetic acid.

27.59 The mechanical energy generated from the translocation of protons in oxidative phosphorylation is first displayed in the rotating ion channel of ATPase.

27.61 Citrate and malate, intermediates in the citric acid cycle, both have carboxyl groups and a hydroxyl group.

27.63 Myosin, the thick filament in muscle, is an enzyme that acts as an ATPase.

27.65 Isocitrate has two stereocenters.

27.67 The proton channel is located in the F_0 unit of ATPase that is embedded in the inner membrane.

27.69   All the sources of energy used for ATP synthesis are not completely known at this time. Most of the energy comes from proton translocation. Some energy for proton pumping comes also from breaking the covalent bond of oxygen (reduction of oxygen to water).

27.71   Succinate dehydrogenase catalyzes the oxidation of succinate to fumarate. FAD becomes reduced in the reaction and then transfers its electrons directly to the electron transport chain at complex II.

27.73   The two decarboxylation steps of the citric acid cycle are the source of the carbon dioxide we exhale.

27.75   Since citrate plays a central role in the citric acid cycle, it can be considered a nutrient.

27.77   No, only complexes I, III, and IV generate enough energy to produce ATP.

27.79   Instead of shunting products of sugar metabolism toward the central metabolic pathways like normal cells, cancer cells use the products for uncontrolled cell growth.

27.81   Yes, when animals are awake they are more metabolically active than when they are asleep.

27.83   Anabolic pathways that build larger molecules from smaller intermediates are reducing pathways because electrons must be added to construct the chemical bonds necessary for creating larger molecules.

27.85   Mitochondrial processes require oxygen (recall that O_2 is the final electron acceptor in electron transport) to produce ATP. When an athlete sprints, they only need to produce a relatively brief burst of muscle contraction that relies on other mechanisms beside the central pathway to produce ATP.

27.87   Citrate isomerizes to isocitrate to convert a tertiary alcohol to a secondary alcohol. Tertiary alcohols cannot be oxidized, but secondary alcohols can be oxidized to produce a keto group, which is necessary to continue the pathway.

27.89   Iron is found in iron-sulfur clusters in proteins and is also part of the heme group of cytochromes.

27.91   Mobile electron carriers transfer electrons on their path from one large, less mobile, protein complex to another.

27.93 ATP and reducing agents such as NADH and $FADH_2$, generated by the citric acid cycle, are needed for biosynthetic pathways. Also, many of the intermediates of the citric acid cycle are drawn off as part of biosynthetic pathways.

27.95 Biosynthetic pathways are likely to be ones of reduction because their net effect is to reverse catabolism, which is oxidative.

27.97 ATP is not stored in the body. It is hydrolyzed to provide energy for many different kinds of processes and thus turns over rapidly. While the body produces kilograms of ATP every day, it also uses kilograms every day. The average life span of an individual molecule of ATP is less than a second.

27.99 The citric acid cycle generates NADH and $FADH_2$, which are linked to oxygen by the electron transport chain.

Chapter 28: Specific Catabolic Pathways: Carbohydrate, Lipid, and Protein Metabolism

28.1 According to Table 28.2 the ATP yield from stearic acid is 146 ATP. This makes $146/18 = 8.1$ ATP/carbon atom.

For lauric acid (C_{12}):

Step *1* Activation	-2 ATP
Step *2* Dehydrogenation five times	10 ATP
Step *3* Dehydrogenation five times	15 ATP
Six C_2 fragments in common pathway	72 ATP
Total	95 ATP

$95/12 = 7.9$ ATP per carbon atom for lauric acid. Thus stearic acid yields more ATP/C atom. This will be generally true for all the fatty acids. The longer the fatty acid, the higher the ATP per carbon atom as the initial input of -2 ATP is a constant for the process.

28.3 The major use of amino acids is in the synthesis of proteins. Proteins from ingested food are hydrolyzed and the amino acids are used to rebuild proteins that the body constantly degrades. We cannot store amino acids so we need a constant supply in our diet.

28.5 The step referred to is # 4, the aldolase-catalyzed cleavage of fructose 1,6-bisphosphate to glyceraldehyde 3-phosphate and dihydroxyacetone phosphate. Glyceraldehyde 3-phosphate metabolism continues immediately in glycolysis (Step # 5), but the dihydroxyacetone phosphate must first be isomerized to glyceraldehyde 3-phosphate by an isomerase. Glyceraldehyde 3-phosphate and dihydroxyacetone phosphate are in equilibrium and removal of glyceraldehyde 3-phosphate by glycolysis drives the isomerization reaction.

28.7 (a) The steps in glycolysis that need ATP are # 1, phosphorylation of glucose and # 3, the phosphorylation of fructose 6-phosphate.
(b) The steps that yield ATP are # 6, catalyzed by phosphoglycerate kinase and #9, catalyzed by pyruvate kinase.

28.9 ATP is a negative modulator for the allosteric, regulatory enzyme phosphofructokinase, Step # 3, as well as for the enzyme pyruvate kinase, step #9.

28.11 The oxidation of glucose 6-phosphate by the pentose phosphate pathway produces NADPH. This reduced cofactor is necessary for many biosynthetic pathways, but especially for the synthesis of essential fatty acids, and as a defense against oxidative damage.

28.13 The anaerobic degradation of a mole of glucose leads to two moles of lactate. Therefore, three moles of glucose produce six moles of lactate.

28.15 Using the data in Table 28.1, a net yield of 2 ATPs are directly produced by the glycolysis of glucose (glucose to pyruvate). There is the initial expenditure of two ATPs in the first three steps of glycolysis. Then steps 6 and 9 produce 4 ATPs, for a net yield of 2. Most of the ATP from glucose degradation comes from oxidation of the reduced cofactors, NADH and $FADH_2$, linked to respiration and the common pathway.

28.17 (a) Fructose catabolism by glycolysis in the liver yields two ATPs just like glucose.
(b) Glycolytic breakdown of fructose in muscle also yields two ATPs per fructose.

28.19 Enzymes that catalyze the phosphorylation of substrates using ATP are called kinases. Therefore, the enzyme that transforms glycerol to glycerol 1-phosphate is called glycerol kinase.

28.21 (a) The enzymes are thiokinase and thiolase.
(b) "Thio" refers to the presence of the element sulfur.
(c) Both of these enzymes use Coenzyme A that contains a reactive thiol group, –SH, as a substrate.

28.23 Each turn of fatty acid β-oxidation yields one C-2 fragment (acetyl CoA), one $FADH_2$, and one NADH. Therefore, the total yield from three turns is three acetyl CoA, three $FADH_2$, and three NADH. There is still a six-carbon portion of lauric acid left, hexanoyl CoA.

28.25 Using the data from Table 28.2, the yield from the oxidation of one mole of myristic acid is 112 moles of ATP. The process requires six turns of β oxidation and produces 7 moles of acetyl CoA.

28.27 Under normal conditions, the body preferentially uses glucose as an energy source. When a person is well fed (balance of carbohydrates and fats and proteins), fatty acid oxidation is slowed and the acids are linked to glycerol and are stored in fat cells for use in times of special need. Fatty acid oxidation becomes important when glucose supplies begin to be depleted, for example, during extensive physical exercise or fasting or starvation.

28.29 The transformation of acetoacetate to β-hydroxybutyrate is a redox reaction using the cofactor, NADH. Acetone is produced by the spontaneous decarboxylation of acetoacetate.

28.31 Oxaloacetate produced from the carboxylation of PEP normally enters the citric acid cycle at Step 1. As we will learn in the next chapter, oxaloacetate may also be used to synthesize glucose.

28.33 Oxidative deamination of alanine:

$$CH_3-\underset{\underset{NH_3^+}{|}}{CH}-COO^- + NAD^+ + H_2O \longrightarrow CH_3-\underset{\underset{O}{||}}{C}-COO^- + NADH + H^+ + NH_4^+$$

$$\qquad\qquad\text{Alanine}\qquad\qquad\qquad\qquad\qquad\qquad\text{Pyruvate}$$

28.35 One of the nitrogen atoms in urea comes originally from an ammonium ion through the intermediate carbamoyl phosphate (steps 1 and 2 in the urea cycle). The ammonium ion was probably released from an amino acid by oxidative deamination. The other nitrogen atom of urea comes from aspartate that enters the urea cycle at step 3.

28.37 (a) The toxic product from the oxidative deamination of Glu is the ammonium ion.
(b) The ammonium ion is converted to urea by the urea cycle and eliminated in the urine.

28.39 Tyrosine is considered a glucogenic amino acid because pyruvate can be converted to glucose when the body needs it. Any amino acid with an easy pathway to pyruvate will be considered glucogenic.

28.41 During initial hemoglobin catabolism, the heme group and globin proteins are separated. The globins are hydrolyzed to free amino acids that are recycled and the iron is removed from the porphyrin ring and saved in the iron-storage protein, ferritin, for later use.

28.43 During exercise, normal glucose catabolism shifts to a greater production of lactate rather than conversion of pyruvate to acetyl CoA and entry into the citric acid cycle. This shift in metabolism is a result of a depletion of oxygen supplies. A build-up of lactate in muscle leads to a lowering of pH which effects myosin and actin action ultimately leading to cramps.

28.45 The acidic nature of the ketone bodies lowers blood pH. This increase in proton concentration is neutralized by the bicarbonate/carbonic acid buffer system present in blood (Section 9.11D and Chemical Connections 9D).

28.47 Protein turnover is an important process because there is no storage form of proteins, thus for synthesis of new proteins, old proteins must be degraded.

28.49 Ubiquitin is a small 76 amino acid protein that is added to other proteins to mark them for degradation by the proteasome.

28.51 When phenylalanine accumulates, it is converted to phenylpyruvate via transamination:

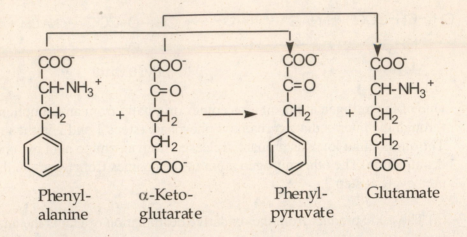

| Phenyl- | α-Keto- | Phenyl- | Glutamate |
| alanine | glutarate | pyruvate | |

28.53 The color changes observed in bruises represent the redox reactions in heme degradation. Black and blue colors are due to congealed blood, green to the biliverdin, and yellow to bilirubin (Figure 28.10).

28.55 The production of ethanol from glucose in yeast, called anaerobic glycolysis, is similar to the conversion of glucose to lactate in humans. There is no net production of reduced coenzymes for recycling. Only two moles of ATP are produced in the conversion of one mole of glucose to ethanol.

28.57 The pentose phosphate pathway is used to produce ribose from glucose. Glucose 6-phosphate entering the pathway is oxidized to form ribulose 5-phosphate and NADPH. Ribulose 5-phosphate can then be converted to ribose 5-phosphate and then to ribose.

28.59 Step 9 in glycolysis, PEP + ADP react to form pyruvate + ATP, confirms that PEP has more energy than ATP. Otherwise there would not be enough energy to drive the reaction toward formation of the product ATP.

28.61 Structure of carbamoyl phosphate:

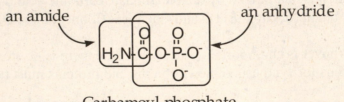

Carbamoyl phosphate

28.63   Phosphoenolpyruvate may be converted to pyruvate (pyruvate kinase) and the pyruvate carboxylated to oxaloacetate.

28.65   Glycolysis is considered oxidative because in addition to reaction 5 (see Figure 28.4) where NADH is formed, there are no reduction reactions, thus overall it is considered oxidative. In addition, pyruvate must be oxidatively decarboxylated to form acetyl CoA which enters the citric acid cycle.

28.67   When fats are catabolized, glycerol is released which then enters glycolysis after being converted in 2 steps to dihydroxyacetone phosphate (See Figure 28.4 and Section 28.4), an intermediate of glycolysis.

28.69   Table 28.1 refers to the complete oxidation of glucose to carbon dioxide and water. Glycolysis is only the beginning of the complete process.

28.71   The conversion of pyruvate to lactate regenerates NAD^+ in anaerobic metabolism. Lactate is a dead end metabolite in the muscle, but it can be sent to the liver in the bloodstream and reconverted to glucose.

28.73   Amino acids yield citric acid cycle intermediates on degradation, thus providing energy.

28.75

Catabolism	Anabolism
Oxidative	Reductive
Energy-yielding	Energy-requiring

28.77   In photosynthesis, carbon dioxide and water combine to produce glucose and oxygen. In aerobic metabolism, glucose and oxygen combine to form carbon dioxide and water. On that level, the two processes are the exact opposite of one another. However, photosynthesis requires light energy from the sun, and glucose breakdown yields energy. The overall pathways are also quite different with respect to the intermediate steps.

28.79   Fats require more oxidation steps to produce carbon dioxide and water than is the case with carbohydrates. The oxidation reactions yield energy in the form of ATP, so fats provide more energy than carbohydrates on a per carbon atom basis.

28.81   The phosphate group on glycolytic intermediates is charged. Since glycolysis takes place in the cytoplasm, the charge is an advantage because it will make these compounds less likely to pass through the cell membrane. The mitochondrion is surrounded by a double membrane, so its contents are less likely to leak out.

28.83   The figure does indicate that ATP is produced. In the center of the citric acid cycle in the mitochondrion shown, it is indicated that NADH and $FADH_2$ are formed. The reoxidation of these reduced coenzymes leads to ATP. While not shown explicitly, the β-oxidation step shown in the top of the mitochondrion also produces ATP via production of these reduced coenzymes.

Chapter 29: Biosynthetic Pathways

29.1 Your text states in Section 29.1 several reasons why anabolic pathways are different from catabolic: (1) Duplication of pathways adds flexibility. If the normal biosynthetic pathway is blocked, the body can use the reverse of the catabolic pathway to make the necessary metabolites. (2) Separate pathways allow the body to overcome the control of reactions by reactant concentration (Le Chatelier's principle). (3) Different pathways provide for separate regulation of each pathway. Although there are many differences between anabolism and catabolism, we will also note similarities that allow for coordinated regulation and proper balancing of concentrations.

29.3 The cellular concentration of inorganic phosphate, a reagent used in phosphorylation reactions, is very high; therefore, the reaction is driven in the direction of glycogen breakdown. In order to ensure the presence of glycogen when needed, it must have an alternate synthetic pathway.

29.5 The major difference between the overall reactions of photosynthesis and respiration is the direction of the reactions. They are the reverse of each other:

$$6CO_2 + 6H_2O \rightarrow C_6H_{12}O_6 + 6O_2 \quad \text{photosynthesis}$$
$$C_6H_{12}O_6 + 6O_2 \rightarrow 6CO_2 + 6H_2O \quad \text{respiration}$$

29.7 A compound that can be used for gluconeogenesis:
(a) From glycolysis: pyruvate
(b) From the citric acid cycle: oxaloacetate
(c) From amino acid oxidation: alanine

29.9 The brain obtains most of its energy from glucose that is supplied by the blood. The brain has little or no capacity to store glucose in glycogen.
(a) Glycolysis outside of the brain is not used as this would use up glucose. Glycolysis is used in the brain as the initial way of utilizing glucose for energy. The glucose used would be metabolized aerobically.
(b) gluconeogenesis is used outside of the brain to produce glucose during starvation.
(c) glycogenesis is the production of glycogen, which is not relevant to this question. Brain cells are also able to obtain some energy from degradation of ketone bodies. Thus the primary sources of energy for the brain during starvation will be ketone bodies and glucose produced by gluconeogenesis in the liver.

29.11 Maltose is a disaccharide that is composed of two glucose units linked by an α-1,4-glycosidic bond (Section 20.4C). We know that in glycogen synthesis, the UDP-glucose can combine with another glucose to add to the glycogen chain. Therefore, we could envision a similar reaction to make maltose. The enzyme might be called maltose synthase:

$$\text{UDP-glucose} + \text{glucose} \longrightarrow \text{maltose} + \text{UDP}$$

29.13 Uridine triphosphate (UTP) is a nucleoside triphosphate similar to ATP. The constituents are: a nitrogen base, uracil; a sugar, ribose; and three phosphates.

29.15 (a) The biosynthesis of fatty acids occurs primarily in the cell cytoplasm. Here acetyl CoA is used to make palmitoyl CoA. Extension of the carbon chain to stearate and desaturation to form carbon-carbon double bonds occurs in mitochondria and the endoplasmic reticulum.
(b) Fatty acid catabolism does not occur in the same location as anabolism. The enzymes for β-oxidation are located in the mitochondrial matrix.

29.17 In fatty acid synthesis, the compound that is added repeatedly to the enzyme, fatty acid synthase, is malonyl CoA, which has a three-carbon chain.

29.19 The carbon dioxide is released from malonyl-ACP which leads to the addition of two carbons to the growing fatty acid chain.

29.21 If one considers only what is happening to the fatty acid, removal of two hydrogens and two electrons, then it looks like oxidation only. However, the reaction is much more complex and involves a cofactor, NADPH and the substrate, oxygen. Both the fatty acid and NADPH undergo two-electron oxidation. The four electrons and protons are used to reduce oxygen to water:

$$O_2 + 4H^+ + 4e^- \longrightarrow 2H_2O$$

29.23 The only structural difference between NADH and NADPH is a phosphate group on one of the ribose units of NADPH. When considering the binding of NADPH to an NADH-requiring enzyme, two factors are important—size and charge. The phosphate makes the NADPH bulky and the NADH binding site may not be able to accommodate the larger size of the cofactor. In terms of charge, NADPH has two negative charges not present in NADH. The NADH binding site may have amino acid residues that have negatively-charged side chains like Glu or Asp. These would repel NADPH, but could hydrogen bond to the free hydroxyl group in NADH.

29.25 Humans have enzymes that catalyze the oxidation of saturated fatty acids to mono-unsaturated fatty acids with the double bond between carbons 9 and 10. For example, we can make palmitoleic acid from palmitic acid and oleic acid from stearic acid. Humans do not have enzymes that introduce a double bond beyond the 10th carbon. Therefore humans cannot make linoleic (double bonds at carbons 9-10 and 12-13) or linolenic acid (double bonds at carbons 9-10, 12-13, and 15-16). Those fatty acids are essential in the diet.

29.27 To make a glucoceramide, sphingosine is reacted with an acyl CoA that adds a fatty acid in amide linkage. Glucose is added to the hydroxyl group of sphingosine using the activated form, UDP-glucose (see Section 21.8).

29.29 All of the carbons in cholesterol orginate in acetyl CoA. An important intermediate in the synthesis of the steroid is a C-5 fragment called isopentenyl pyrophosphate:

$$3\text{AcetylCoA} \longrightarrow \text{mevalonate} \longrightarrow \text{isopentenyl pyrophosphate} + CO_2$$

C-2   C-6   C-5

29.31 An amino acid is synthesized by the reverse of oxidative deamination (Section 29.5). The amino acid product is aspartate. NADH will be oxidized to NAD^+.

29.33 The products of the transamination reaction shown are valine and α-ketoglutarate.

$$(CH_3)_2CH\text{-}\overset{O}{\overset{\|}{C}}\text{-}COO^- + {}^-OOC\text{-}CH_2\text{-}CH_2\text{-}\underset{\underset{NH_3^+}{|}}{CH}\text{-}COO^- \longrightarrow$$

The keto form of valine   Glutamate

$$(CH_3)_2CH\text{-}\underset{\underset{NH_3^+}{|}}{CH}\text{-}COO^- + {}^-OOC\text{-}CH_2\text{-}CH_2\text{-}\overset{O}{\overset{\|}{C}}\text{-}COO^-$$

Valine   α-Ketoglutarate

29.35 The carbon dioxide that is used to make carbohydrates in plants is reduced by the cofactor NADPH.

29.37 An enzyme currently under study is Acetyl-CoA Carboyxlase 2 (ACC2), one form of the enzyme that makes malonyl-CoA. This enzyme found in cardiac and skeletal muscle affects fat metabolism. Scientists believe that if they could inhibit ACC2, it would stimulate fatty acid oxidation and inhibit fatty acid synthesis.

29.39 High glucose consumption leads to increased levels of citrate in the cytosol with subsequent conversion to acetyl-CoA by the enzyme ATP-citrate lyase. The acetyl-CoA enters the nucleus and becomes the substrate for enzymes that acetylate histones (histone acetyltransferases). When histones are either acetylated or deacetylated gene expression is altered.

29.41 Yes, the most successful way to lose weight and maintain that reduced weight requires behavior modification. There are definite connections between environment, the nervous system, and over-eating behavior.

29.43 The bonds that connect the nitrogen bases to the ribose units are β-N-glycosidic bonds just like those found in nucleotides.

29.45 The amino acid produced by transfer of the amino group is Phe.

29.47 The structure of a lecithin (also phosphatidyl choline) is in Section 21.6. Its synthesis requires activated glycerol, two activated fatty acids, and activated choline. Since each activation requires one ATP, the total number of ATP molecules needed is four.

29.49 The compound that reacts with Glu in a transamination reaction to form serine is 3-hydroxypyruvate. The reaction is shown below:

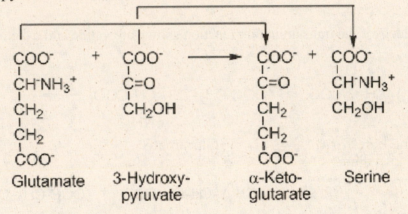

29.51 HMG-CoA is 3-hydroxy 3-methylglutaryl-CoA. Its structure is shown in Section 29.4. Carbon 1 is the carbonyl group linked to the thiol group of CoA.

29.53 Heme is a porphyrin ring with an iron ion at the center. Chlorophyll is a porphyrin ring with a magnesium ion at the center.

29.55 There are several differences, although the principal difference is the starting materials. In plants, sunlight, CO_2 and water are used to form glucose. In animals, ATP and a variety of other starting materials including pyruvate, oxaloacetate, citrate and several amino acids are used to make glucose. The energy needed for glucose synthesis in plants comes from the sun through the process of photosynthesis, whereas in animals ATP is directly used in a process called gluconeogenesis. In plants, photosynthesis is a 2 stage process that first generates ATP and NADPH in a process called the light reactions that take place in chloroplasts. Then in a second process called the dark reactions the ATP along with NADPH is used to reduce CO_2 to form glucose. Animals use a process very similar to the reverse of glycolysis using many of the same glycolytic enzymes, but also a few unique ones to synthesize glucose from many starting materials.

29.57   Fatty acid biosynthesis takes place in the cytoplasm, requires NADPH, and uses malonyl-CoA. Fatty acid catabolism takes place in the mitochondrial matrix, produces NADH and $FADH_2$, and has no requirement for malonyl-CoA.

29.59   Photosynthesis has high requirements for light energy from the sun.

29.61   Lack of essential amino acids would hinder the synthesis of the protein part. Gluconeogenesis can produce sugars even under starvation conditions.

29.63   Separation of catabolic and anabolic pathways allows for greater efficiency, especially in control of the pathways. It allows for both processes to be going on simultaneously in the body as conditions can be different in different cell compartments.

29.65   It will not be useful to give the rats the missing amino acid later, as the other amino acids will have been metabolized. To make necessary proteins, the body needs to have all of the essential amino acids available.

Chapter 30: Nutrition

30.1 No, nutrient requirements vary from person to person. The recommended daily allowances (RDA) stated by the government are average values and they assume a wide range of needs in different individuals. Some of the factors that change dietary needs include age, level of activity, genetics, and geographical region of residence.

30.3 Sodium benzoate is not catabolized by the body; therefore, it does not comply with the definition of a nutrient—components of food that provide growth, replacement, and energy. Calcium propionate enters mainstream metabolism by conversion to succinyl CoA and catabolism by the citric acid cycle.

30.5 The Nutrition Facts label found on all foods must list the percentage of daily values for four important nutrients: vitamins A and C, calcium, and iron.

30.7 Fiber is an important non-nutrient found in some foods. It is the indigestible portion of fruits, vegetables, and grains. Chemically, fiber is cellulose, a polysaccharide that cannot be degraded by humans. It is important for proper operation of dietary processes, especially in the colon.

30.9 The basal caloric requirement is calculated assuming the body is completely at rest. Because most of us perform some activity, we need more calories than this basic minimum. The intake of 2100 Cal/day for a young woman is a peak requirement. The difference between the basic level and the peak is needed to produce energy for activity.

30.11 Assume that each pound of body fat is equivalent to 3500 Cal. Therefore, the total number of calories that must be deleted from the diet is 3500 Cal/lb x 20 lb = 70,000 Cal. Because this must be done in 60 days, the amount restricted each day is 70,000 Cal ÷ 60 days = 1167 Cal/day. The caloric intake each day should be 3000 − 1167 = 1833 Cal, or approximately 1800 Cal/day.

30.13 Water makes up about 60 % of our body weight; therefore, one might assume she/he can lose weight by taking diuretics to enhance water release. However, this would only serve as a "quick fix" as the weight would return rapidly. In addition, it could be dangerous because the body needs a constant supply of water. The only way to effectively lose

30.15 Amylose is a storage polysaccharide that is one of the components in starch. Chemically it is an unbranched polysaccharide composed of glucose residues linked by α-1,4-glycosidic bonds. α-Amylase is an enzyme that catalyzes the hydrolysis of the glycosidic bonds at random sites in amylose. Thus the product would be different-sized, oligosaccharide fragments much smaller than the original amylose molecules.

30.17   No. Dietary maltose, the disaccharide composed of glucose units linked by an α-1,4-glycosidic bond, is rapidly hydrolyzed in the stomach and small intestines. By the time it reaches the blood, it is the monosaccharide glucose.

30.19   Arachidonic acid is produced from the essential, unsaturated fatty acid, linoleic acid.

30.21   No, lipases degrade neither cholesterol nor fatty acids. Lipases catalyze the hydrolysis of the ester bonds in triglycerides, releasing free fatty acids and glycerol.

30.23   Yes, it is possible for a vegetarian to obtain a sufficient supply of adequate proteins; however, the person must be very knowledgeable about the amino acid content of vegetables. It is very difficult to find a single vegetable that has complete proteins, that is, proteins with every essential amino acid. It is important for a vegetarian to eat a variety of vegetables in a proper balance so that all of the essential amino acids, fats, and nutrients are present.

30.25   Dietary proteins begin degradation in the stomach that contains HCl in a concentration of about 0.5 %. Trypsin is a protease present in the small intestines that continues protein digestion after the stomach. Stomach HCl denatures dietary protein and causes somewhat random hydrolysis of the amide bonds in the protein. Fragments of the protein are produced. Trypsin catalyzes hydrolysis of peptide bonds only on the carboxyl side of the amino acids Arg and Lys.

30.27   There is no place to store excess protein consumed in the diet unlike carbohydrate conversion to glycogen and fat storage in adipose tissue. Excess consumed protein is converted to fat. Because of this, an adequate amount of protein must be consumed daily.

30.29   The rice/water diet provides sufficient calories for the basal caloric requirement; however, the diet is lacking important nutrients such as some essential amino acids (e.g.,Thr, Lys), essential fats, vitamins, and minerals. It is expected that many of the prisoners will develop deficiency diseases in the near future.

30.31   Limes provided sailors with a supply of vitamin C to prevent scurvy.

30.33   Vitamin K is essential for proper blood clotting.

30.35   The only disease that has been proven scientifically to be prevented by vitamin C is scurvy.

30.37   Vitamins E and C, and the carotenoids may have significant effects on respiratory health. This may be due to their activity as antioxidants.

30.39   There is a sulfur atom in biotin and in vitamin B-1 (also called thiamine).

30.41   Vitamin A is fat soluble which means that it can be stored long term in fat tissue. When taken in large doses, vitamin A is especially toxic.

30.43   The original food pyramid did not consider the difference between types of nutrients. It assumed that all fats were to be limited and that all carbohydrates were healthy. The new guide recognizes that polyunsaturated fats are necessary and that carbohydrates from whole grains are better for you than those from refined sources.

30.45   All proteins, carbohydrates, and fats in excess have metabolic pathways that lead to increased levels of fatty acids. However, there is no pathway that allows fats to generate a net surplus of carbohydrates. Thus, fat stores cannot be used to make carbohydrates when a person's blood glucose is low. This reality has a lot to do with the difficulty of diets.

30.47   All effective weight loss is based upon increasing activity while limiting caloric intake. However, it is more effective to concentrate on increasing activity than limiting intake. While it is difficult to lose weight by dieting because of the fact that fats cannot be used to make carbohydrates, exercise uses up fatty acids quickly without necessarily depleting glycogen stores. This makes exercise a more efficient way to lose weight.

30.49   Theoretically speaking, if humans had the glyoxylate pathway, they would be able to convert fat into glucose. It is the two decarboxylation steps of the citric acid cycle that makes it impossible to convert fat to glucose. If these decarboxylations could be bypassed, then we could convert fat to glucose. If we could convert fat to glucose, then dieting would be easy and weight loss would be quick and effective.

30.51   Iron is required because there are many proteins in the body that have iron as part of their structures. The most obvious one we have studied is hemoglobin, but there are many others.

30.53   Factors the affect absorption include the solubility of a given iron-containing compound, the presence of antacids in the digestive tract, and the source of the iron.

30.55   Note the structure of creatine in Chemical Connections 30D. The amino acid that most resembles the structure of creatine is arginine. The top part of creatine, the carbon bound to three nitrogens, is the same as the side chain of Arg. Also, both have carboxylate groups at the other end of the molecule.

30.57   Carbohydrates should be consumed before the event as part of a high-carbohydrate diet. This is still considered the best diet for athletes. For competitions lasting longer than about an hour, carbohydrates should also be consumed during the event to prevent the complete depletion of glycogen and subsequent lowering of blood glucose levels.

30.59   Caffeine acts as a central nervous system stimulant, which provides a feeling of energy that athletes often enjoy. In addition, caffeine reduces insulin levels and stimulates oxidation of fatty acids, which would be beneficial to endurance athletes.

30.61   The major considerations are cost vs. perceived value. Organic food can be 100% more expensive than non-organic food, making it out of reach of many people. The data are far from conclusive regarding the health benefits of organic food adding to the dilemma of deciding the cost to value. Another important consideration is the type of food being considered when attempting to ascertain the importance of the food being organic. An example is the difference between a pesticide that would be concentrated in a banana peel vs. one that would concentrate in the banana itself. If the pesticide is concentrated in the banana peel, then it might not be worth spending extra money to buy organic bananas, whereas it would be worth the money if the pesticide is concentrated in the fruit. Another important consideration is who the consumer is. Pesticides are more dangerous for children and pregnant women than for others.

30.63   A variety of animal produced (endocannabinoids), plant produced – THC, and synthetic cannabinoids like HU-210 can act as ligands and stimulate these receptors.

30.65   A study of 600 depression cases showed that those who consumed diets rich in trans fats were almost 50% more likely to be depressed.

30.67   Several studies have showed that both men and women who have vitamin B_{12} deficiencies are much more likely to be depressed. This could involve the molecule SAMe (*S*-adenosylmethionine) a methyl-group donor that is important for the metabolism of neurotransmitters linked with depression. Vitamin B_{12} along with folic acid are necessary for the synthesis of SAMe.

30.69   The vitamin pantothenic acid is part of Coenzyme A.
         (a) Glycolysis: pyruvate dehydrogenase in Reaction 12, Figure 28.4, uses Coenzyme A.
         (b) Fatty acid synthesis: the first step that involves the enzyme fatty acid synthase.

30.71   Proteins that are ingested in the diet are degraded to free amino acids that are then used to build proteins that carry out specific functions. Two very important functions include structural integrity and biological catalysis. Our proteins are constantly being turned over, that is, continuously being degraded and rebuilt using free amino acids.

30.73   The very tip of the food pyramid displays fats, oils, and sweets, with the cautionary statement, "Use sparingly". We can omit sweets, meaning refined sugars, completely from the diet; however, complete omission of fats and oils is dangerous. We must have dietary fats and oils that contain the two essential fatty acids, linoleic acid and linolenic acid. It may be possible that the essential fatty acids are present as components in other food groups, i.e. the meat, poultry, fish group.

30.75 Walnuts are not just a tasty snack, they are a healthy one. Walnuts have protein, in fact, nuts are included in a group of the Dept. of Agriculture's food pyramid. Walnuts are also a good source of vitamins and minerals including vitamins E, B, biotin, potassium, phosphorus, zinc, manganese, and others.

30.77 No, the lecithin is degraded in the stomach and intestines long before it could get into the blood. The phosphoglyceride is degraded to fatty acids, glycerol, and choline, that are absorbed through the intestinal walls.

30.79 Patients who have had ulcer surgery are administered digestive enzymes that may have been lost during the procedure. The enzyme supplement should contain proteases to help break down proteins, and lipases to assist in fat digestion.

Chapter 31: Immunochemistry

31.1   Examples of external innate immunity include action by the skin, tears, and mucus, all of which are barriers to entrance of foreign particles.

31.3   The skin fights infection by providing a barrier against penetration of pathogens. The skin also secretes lactic acid and fatty acids, both of which create a low pH thus inhibiting bacterial growth.

31.5   Innate immunity processes have little ability to change in response to immune dangers. The key features of adaptive (acquired) immunity are specificity and memory. The acquired immune system uses antibody molecules that are very specific for each type of invader. In a second encounter with the same danger, the response is more rapid and more prolonged than the first.

31.7   T cells originate in the bone marrow, but grow and develop in the thymus gland. B cells originate and grow in the bone marrow.

31.9   Macrophages are the first cells in the blood that encounter potential threats to the system. They attack virtually anything that is not recognized as part of the body including pathogens, cancer cells, and damaged tissue. Macrophages engulf an invading bacteria or virus and kill it using nitric oxide (see section 31.2B).

31.11 T cells recognize only peptide/protein antigens. This is in contrast to B cells which can recognize antigens from other molecule types, such as complex carbohydrates.

31.13 Class II MHC molecules pick up damaged antigens. A targeted antigen is first processed in lysosomes where it is degraded by proteolytic enzymes. An enzyme, GILT, reduces the disulfide bridges of the antigen. The reduced peptide antigens unfold and are further degraded by proteases. The peptide fragments remaining serve as epitopes that are recognized by class II MHC molecules (Figure 31.8).

31.15 MHC molecules are transmembrane proteins that belong to the immunoglobulin superfamily. They have peptide-binding variable domains. Class I molecules are single-chain polypeptides, whereas class II are protein dimers. They are originally present inside cells until they become associated with antigens and move to the surface membrane.

31.17 If we assume that the rabbit has never been exposed to the antigen, the response will be 1-2 weeks after the injection of antigen.

31.19 (a) IgE molecules have a carbohydrate content of 10-12% which is equal to that of the IgM molecules. IgE molecules have the lowest concentration in the blood. The blood concentration of IgE is about 0.01-0.1 mg/100mL of blood.

239

(b) IgE molecules are attached to basophils and mast cells. When they encounter their antigens, the cells release histamines that cause the effects of hay fever and other allergies. They are also involved in the body's defense against parasites.

31.21 The two Fab fragments would be able to bind an antigen. Note in Figure 31.9 that these fragments contain the variable protein sequence regions and hence are able to be changed during synthesis against a specific antigen.

31.23 Immunoglobulin superfamily refers to all of the proteins that have the standard structure of a heavy chain and a light chain. In terms of amino acid sequence, each molecule has a constant region and a variable region. Molecules in the immunoglobulin super-family are the Ig molecules themselves (e.g., IgG, IgM, IgE), T-cell receptors, and the MHCs.

31.25 Antibodies and antigens are held together by weak, noncovalent interactions. These are: hydrogen bonds, electrostatic interactions (dipole-dipole), and hydrophobic interactions. We have encountered these kinds of interactions also in enzyme-substrate complexes and hormone-receptor complexes.

31.27 Antibody diversity is caused by several factors. There are two major types of light chains, kappa and lambda, which can combine with heavy chains. Each type of chain shows the recombination of the V, J, and D genes discussed in 31.4D, when the B and T cells are differentiating. In addition, mutations occur as the cells are maturing, causing another level of diversity.

31.29 T cells carry on their surfaces unique receptor proteins that are specific for antigens. These receptors (TcR), members of the immunoglobulin superfamily, have constant and variable regions. They are anchored in the T cell membrane by hydrophobic interactions. They are not able to bind antigens alone, but they need additional protein molecules called cluster determinants that act as coreceptors. When TcR molecules combine with cluster determinant proteins, they form T-cell receptor complexes (TcR complex).

31.31 The components of the T cell receptor complexes are (1) accessory protein molecules called cluster determinants and (2) the T-cell receptor.

31.33 The adhesion molecule in the TcR complex that assists HIV infection is the cluster determinant 4 (CD4).

31.35 The CD4 and CD8 molecules act as adhesion molecules, helping the T cell receptor bind to the antigens and to help the T cell dock with the antigen presenting cell (APC) or B cell.

31.37 Cytokines are glycoproteins that interact with cytokine receptors on macrophages, B cells, and T cells. They do not recognize and bind antigens.

31.39 Chemokines are a class of cytokines that send messages between cells. They attract leukocytes to the site of injury and bind to specific receptors on the leukocytes.

31.41 Cytokines are classified in several ways, but the main way is via their secondary structures. Cytokines such as interleukin-2 contain four helical segments. Another type such as tumor necrosis factor have only β-pleated sheets for a secondary structure. Another one, such as epidermal growth factor has a combination of α-helix and β-pleated sheet. A subgroup of cytokines, called chemokines, have 4 cysteine residues.

31.43 Jenner purposefully infected a boy with cowpox to test whether it would protect the boy against a closely related virus, smallpox. What made this experiment especially unethical and certainly illegal if done today, was that after infecting the boy with cowpox, a mild disease, Jenner injected the boy with a lethal dose of smallpox. If the vaccine had not worked, the boy would have died.

31.45 The word vaccination was coined by the French who intended it be a derisive term. It literally means "encowment."

31.47 Subunit vaccines are the safest, since only a small portion of the pathogen is present; there is no possibility of pathogen replication. The other vaccine types involve injecting the whole pathogen into the patient.

31.49 Dendritic cells encounter the vaccine in the tissue and engulf it. They then become activated, release cytokines to stimulate the immune system, and travel to the lymph nodes to present antigen to immature T cells. They represent the first step of the immune system, so it is important to engage them when developing an effective vaccine.

31.51 Developing successful vaccines against HIV, tuberculosis, hepatitis C, and malaria has proven very difficult. No successful vaccines exist yet for these pathogens.

31.53 LPS worked because it bound to Toll-like receptor 4 (TLR-4) on dendritic cells and caused the cells to release cytokines.

31.55 The T cells mature in the thymus gland. During maturation those cells that fail to interact with MHC and thus cannot respond to foreign antigens are eliminated by a special selection process. T cells that express receptors (TcR) that may interact with normal-self antigens are eliminated by the same selection process.

31.57 A signaling pathway that controls the maturation of B cells is the phosphorylation pathway activated by tyrosine kinase and deactivated by phosphatase.

31.59   The cytokines and chemokines are thought to play a major role in autoimmune diseases.

31.61   Human Immunodeficiency Virus binds to and enters the helper T cells in humans.

31.63   HIV is difficult to find because it mutates so quickly, making it difficult to create an effective vaccine against the virus. The gp120 protein on the virus changes conformations when it binds to the CD4 molecule, so antibodies elicited against the virus during the active part of the infection will not recognize the virus that is floating free. HIV also has proteins on its surface that block the action of natural killer cells, so it can evade the innate immune system. The virus also cloaks its outer membrane with sugars that are very similar to the natural sugars of the host cell membranes.

31.65   It is difficult to create an effective antibody to a virus that mutates quickly. Even if the antibody is successful at neutralizing most of the virus particles, just a few survivors can repopulate the host very quickly. In addition, most attempts to generate antibodies have led to large numbers of antibodies that were not effective.

31.67   A "two-in-one" antibody unlike normal antibodies recognizes two different epitopes. This makes it especially useful for a rapidly mutating virus because two different epitopes can be targeted simultaneously. The likelihood that both targets will have mutated is low, making the "two-in-one" antibody a powerful treatment.

31.69   Monoclonal antibodies can be created to attack very specific proteins that are on the surface of cancer cells. They are very versatile and can be created to attack a wide variety of targets.

31.71   The direct evidence of the importance of HER2 is that when a red fluorescent tag is made to the HER2 protein, it can be shown that many types of cancer cells have much larger quantities of the HER2 protein than normal cells.

31.73   Some cell receptors that can lead to cancer are members of the class of tyrosine kinases. When a cell growth factor binds to the receptor, it initiates a chain of events beginning with the action of the tyrosine kinase. This is one reason that monoclonal antibodies have been designed to neutralize tyrosine kinase.

31.75   Many people are allergic to certain antibiotics, such as penicillin. These allergies can be very potent, even causing death if the person is exposed repeatedly.

31.77   Sex workers in the Philippines were in the habit of using low doses of penicillin to help prevent the spread of STD's. In fact, this practice had the opposite effect of allowing the development of resistant strains of gonorrhea.

31.79 There is a link between strep throat and the disease, rheumatic fever. A particular protein found in the group A streptococci, the M protein, is similar in structure to proteins found in the valves of the heart. In an attempt to fight strep throat, the body makes antibodies to the strep, but these same antibodies can attack the body's own heart valves.

31.81 Stem cells can be transformed into other cell types. Scientists are working to find ways to use stems cells to repair damaged nerve tissue or brain tissue. In some animal models, brain cell function has been restored after a stroke by adding stem cells to the brain in the area of the damage. It is hoped that stem cells will lead to being able to cure spinal cord damage and brain injuries.

31.83 The flu is considered much more dangerous because it killed more than 50 million in 1918 alone. HIV by comparison, has only infected 40 million and it has taken 30 years to get to that point.

31.85 No, it's an incorrect statement. H5N1 (the avian flu) has killed 60% of those infected, although it has only infected a small number of people – a few hundred people worldwide. It is not very transmissible. H1N1 (the swine flu) on the other hand, is very transmissible and has infected far more people and caused far more deaths worldwide. Although, the percentage of those infected with H1N1 who are killed is much lower than H5N1.

31.87 Two teams created monoclonal antibodies against HA that recognized epitopes on the stem portion of HA that were relatively constant. Interestingly, instead of just blocking one type of HA like most antibodies, they blocked 8 of 15 different varieties of HA. A vaccine that caused formation of this type of antibody in humans could be especially effective.

31.89 The IgA molecules, present in mucous membranes, milk, and tears, are the first line of defense against bacteria.

31.91 Chemokines (or more generally cytokines) help leukocytes migrate out of a blood vessel to the site of injury. Cytokines help the proliferation of leukocytes.

31.93 A compound called 12:13 dEpoB, a derivative of epothilon B, is being studied as a anticancer vaccine.

31.95 TNF (tumor necrosis factor) receptors are found on the surface of tumor cells.